Profile of a
CRIMINAL MIND

Profile of a CRIMINAL MIND

How Psychological Profiling Helps Solve True Crimes

Brian Innes

Reader's Digest

The Reader's Digest Association, Inc.
Pleasantville, New York/Montreal

A READER'S DIGEST BOOK

This edition published by the Reader's Digest Association
by arrangement with Amber Books Ltd

Editorial and design by
Amber Books Ltd
Bradley's Close
74–77 White Lion Street
London N1 9PF
United Kingdom
www.amberbooks.co.uk

FOR AMBER BOOKS:
Project Editor: Michael Spilling
Design: Jerry Williams
Picture Research: Natasha Jones
Illustrations: Rowena Dugdale

FOR READER'S DIGEST:
U.S. Project Editor: Susan Randol
Canadian Project Editor: Pamela Johnson
Project Designer: George McKeon
Creative Director: Michele Laseau
Executive Editor, Trade Publishing: Dolores York
Director, Trade Publishing: Christopher T. Reggio
Vice President & Publisher, Trade Publishing: Harold Clarke

Library of Congress Cataloging-in-Publication Data

Innes, Brian.
    Profile of a criminal mind: how psychological profiling helps solve true crimes / Brian Innes.
        p. cm.
    Includes index.
    ISBN 0-7621-0407-4
    1. Criminal investigations—Psychological aspects. 2. Criminal
behavior—Research—Methodology. 3. Criminal methods—Research—Methodology. I. Title.

HV8073.5.I56 2003
363.25'8—dc21

                                                                    2003046569

Address any comments about *Profile of a Criminal Mind* to:
    The Reader's Digest Association, Inc.
    Adult Trade Publishing
    Reader's Digest Road
    Pleasantville, NY 10570-7000

For Reader's Digest products and information, visit our website:
    www.rd.com (in the United States)
    www.readersdigest.ca (in Canada)

Printed in Italy

1 3 5 7 9 10 8 6 4 2

CONTENTS

INTRODUCTION

The criminal has been an unwelcome element of society since time immemorial, and the attempt to penetrate his or her mind, to discover whether he or she differs significantly from the person who is considered an honest citizen – and if so, to what degree – has preoccupied people for centuries. For a long time, most research was superficial, directed solely toward ways of identifying the physical characteristics of known criminals – an approach that was of limited value in the investigation or prevention of crime.

But as interest in the workings of the human mind developed, attention turned, either to ways of identifying the thought processes of criminals and therefore frustrating their future crimes, or toward the possibility of their subsequent reform. It is only within the last hundred years, however, that law enforcement authorities have come to realize that an analysis of the specific behavior of an unidentified subject (UNSUB) can provide clues to his or her physical appearance, age, education, social position, and other factors and so help the investigator narrow down the field of inquiry. And it is only with the present-day ready availability of desktop computers that it has proved possible to handle the vast quantities of data on which to base such an analysis.

This approach was originally named psychological profiling, but it is now generally given the more widely applicable name of "offender profiling" or "behavioral analysis." In the beginning it was principally the domain of psychiatrists and psychologists, whose assessments, based upon their clinical experience, were largely intuitive. In many European countries, notably Britain, it remains substantially so. In the United States, however, the technique has been brought to its present state of development by the Federal Bureau of Investigation, and in Canada by the Royal Canadian Mounted Police.

These two organizations have placed a reliance upon the analysis of computerized data that has now been extended to the use of geographical profiling. In Britain, the police computer systems HOLMES and CATCHEM were established as a result of the failure of conventional data gathering in the case of the "Yorkshire Ripper," and other countries have now also followed the FBI example. In addition, there are many independent practitioners who offer their expertise – often to the investigating officers, but also to the prosecution, or even the defense, when an accused person is brought to trial.

There is controversy over who coined the terms "pyschological" or "offender" profiling, but the first systematized application of the technique undoubtedly came with the establishment of the FBI's Behavioral Sciences Unit in 1972, and the subsequent development of VICAP (Violent Criminal Apprehension Program) in 1984. Because of the FBI's concentration on serial murder, rape, and abduction, most attention has been given to these crimes of physical violence; but, with authorities' growing expertise and increasing facilities, similar techniques are gradually being applied to the investigation of a widening range of other crimes.

The use of behavioral analysis in the hunt for violent criminals – especially serial murderers and rapists – had developed with increasing success over the course of more than 20 years before it attracted widespread public interest with the release of the movie *The Silence of the Lambs* in 1992. Based upon a novel by Thomas Harris, the movie featured the cannibalistic psychopath Dr. Hannibal Lecter, who had already appeared in Harris's earlier work, *Red Dragon,* published in 1981.

In writing *Red Dragon,* Harris had sought the advice of the FBI. He was invited to Quantico, the Bureau Academy in Virginia, and allowed to attend training talks on serial killers given by members of the Behavioral Sciences Unit. The character of "Buffalo Bill," the killer who is tracked down in *The Silence of the Lambs,* is a combination of three real-life murderers who were used as examples in these talks.

When it came to the making of the movie of *The Silence of the Lambs,* the FBI was even more cooperative. They allowed the Academy at Quantico to be used for location shooting, and some scenes even included Bureau personnel in minor roles, or as extras. However, the FBI's procedures in the movie provoked severe criticism, not only because the Bureau would never have used a trainee agent on such an assignment, but also because of various inaccuracies of procedures portrayed.

It is hardly surprising that the success of *The Silence of the Lambs* resulted in several highly popular fictional television series. Unfortunately, these have suggested that psychological profiling and behavioral analysis are almost "magic" in their potential, virtually infallible in hunting down the criminal perpetrator. Intuition and clinical experience can play an important part, but the key to success lies in the painstaking assembly of comparative data. As John Douglas, one of the first members of the Behavioral Sciences Unit, has said of the American TV series *Profiler:* "The show makes it seem like a psychic thing. But it all comes from interviewing a lot of subjects and getting a sense of what they're all about."

Apart from this now widely practiced technique of behavioral analysis, which is also of great value in interrogation and crisis negotiation, there are other approaches that can throw light on the personality and thought processes of the unidentified criminal – "getting a sense of what they're all about." In this book, attention is also devoted to what can be detected in verbal and written communications, by approaches such as psycholinguistics, textual analysis, and the assessment of handwriting. These methods are still in their infancy, compared with behavioral analysis, but they are being adopted by law enforcement agencies.

Over the past 20 years, many professional profilers have been happy to publicize their successes, revealing how their methods can spotlight unidentified offenders and bring them to justice. Yet almost nothing is known of how often the system has failed. Nevertheless, the study of the criminal personality, in all its ramifications, is of great importance. Getting inside the criminal mind is an increasingly powerful tool in the war against crime.

SERVANTS WANTED

THE SEARCH FOR THE CRIMINAL PERSONALITY

For many centuries physicians believed that an individual's physical characteristics would reveal whether or not they had a criminal nature. This early 19th-century cartoon (left) satirizes the possibility that employees might be selected by phrenology – examining the shape of the cranium to gauge the personality.

How is it that some people become criminals, while the majority do not? When the same temptations face all of us, why do certain individuals succumb, while others keep to the narrow path of righteousness?

During many centuries, this question was dismissed, almost out of hand, for the answer seemed obvious: either criminals were born that way, unable to control their antisocial instincts, or they had become possessed by malign beings – evil gods, demons, or even the Devil himself.

The ancient Greek philosophers and physicians looked deeply into the question of emotions, their cause and where they might originate in the human body – but their theories remained largely undeveloped for more than 2,000 years, until the time of Sigmund Freud and his associates. As early as the 6th century B.C., the physician Alcmaeon carried out the first dissection of the human body and decided that the seat of reason lay in the brain; while the philosopher Empedocles suggested that love and hate were the fundamental sources of changes in human behavior.

As long ago as 400 B.C., the famous Greek physician Hippocrates described a range of mental disorders of the type that are recognized today, and he spoke out strongly for the legal rights of the mentally

book), the court appointed a "guardian" to represent the accused.

The famous Roman physician Galen (c.130–201 A.D.) theorized that the human "soul" was situated in the brain, and was divided into two parts: the external, which comprised the five senses; and the internal, which governed imagination, judgment, perception, and movement. However, over the next 1,500 years, Galen's theories were almost completely ignored. The medical profession chose to maintain more primitive explanations for the causes of mental disorder, such as witchcraft or demonic possession.

PHYSIOGNOMY

It was during the 16th century that the idea emerged that it was possible to determine the nature of a person by his external features, such as the forehead, mouth, eyes, teeth, nose, or hair. The study was named "physiognomy" by the Frenchman Barthélemy Coclès, and in his book *Physiognomonia* (1533) he provided many woodcuts to illustrate his points.

Gradually, from the 17th century onward, enlightened Western philosophers began to exert an influence on medical thinking, and it was at this time that the term "psychology" was first used. Nevertheless, although the effect of the brain – not only on behavior, but also on diseases – came increasingly to be recognized, external physical characteristics remained predominant in diagnosis. One theory, which contrived to combine both approaches and caught the popular imagination, was "phrenology."

Franz Joseph Gall (1758–1828) was a fashionable physician in Vienna at the end of the 18th century, and he came to the

A romanticized 19th-century portrait of Hippocrates, eminent physician of ancient Greece.

disturbed. At that time, Athenian law recognized the rights of the mentally ill in civil concerns, but not if they were guilty of serious crimes. The influence of Hippocrates brought about changes in the law: If a person on trial could be shown to be suffering from what he called "paranoia" (a term that will recur, with a more specialized meaning, later in this

conclusion that the brain was made up of 33 "organs," whose position and developed size could be discovered by feeling the external "bumps" of the cranium. There were three classes of organ: those controlling fundamental human characteristics; those governing "sentiments," such as benevolence or mirthfulness; and those of a purely intellectual nature, such as the appreciation of size, or the recognition of cause and effect.

> "At the sight of that skull, I seemed to see, all of a sudden...the problem of the nature of the criminal."
> **Cesare Lombroso**

Among the organs that Gall claimed to have identified – which included one that he famously decided indicated the human desire for "procreation" – were those of murder, theft, and cunning. He and his disciple J. K. Spurzheim (1776–1832) – who later named a further four organs – were forced to leave Austria because their ideas conflicted with current medical opinion, but their theories were welcomed in France, Britain, and the United States. In Edinburgh, Spurzheim publicly dissected a human brain, and indicated the position of the various organs; and in the United States "practical phrenologists" traveled from fair to fair, claiming to cure both mental and physical disease.

Phrenology remained a popular preoccupation throughout the 19th century, but contributed little or nothing to an understanding of the criminal mind – although modern neurological research has in fact revealed areas of the brain that control emotions and behavior. The first important development in criminology came, strangely enough, with a revival of interest in physiognomy.

CRIMINAL MAN

The Italian Cesare Lombroso (1836–1909) made one of the first serious studies of criminality. After serving as an army surgeon in the Austro-Italian war of 1866, he was appointed professor of mental diseases at Pavia. There, he began to carry out a succession of dissections on the brains of patients who had died, in the hope of discovering some structural cause for insanity. In this he was unsuccessful, but in 1870 he learned of the German pathologist Rudolf Virchow, who claimed to have discovered unusual features in the skulls of criminals – features that reflected those of prehistoric mankind, or even of animals.

Lombroso at once began to study the physiognomy of criminals in Italian prisons, and performed an autopsy on the body of an executed brigand, paying particular attention to the skull, in which he found just one small, unusual, physical feature, which resembled that of a rodent:

"At the sight of that skull, I seemed to see, all of a sudden, lighted up as a vast plain under

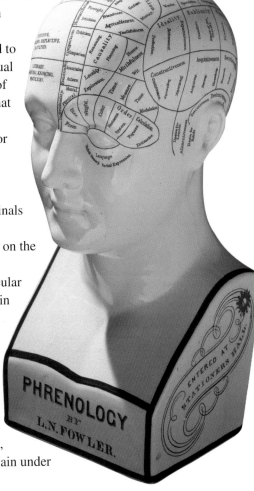

"Reading the bumps" of the human skull, which supposedly revealed the size of specific "organs" of the brain beneath, continued to be of popular interest throughout the 19th century, and well into the 20th. Phrenological heads such as this can still be found in many antique dealers' shops.

a flaming sky, the problem of the nature of the criminal – an atavistic being who reproduces in his person the ferocious instincts of primitive humanity and the inferior animals."

Lombroso's revelation was supported by further studies, and he began to divide his cases into "occasional criminals," who were driven to crime by circumstances, and "born criminals" – those who regularly committed crimes because of some hereditary defect that was apparent in their physical appearance. These "atavistic" individuals were distinguished by their "primitive" features: long arms, acute eyesight (like that of birds of prey), heavy jaws, and "jug" ears.

In 1876, the year in which he was appointed a professor of forensic medicine, Lombroso published his findings in the book *L'Uomo Delinquente* (*Criminal Man*), which quickly achieved international renown. In a later book, *Criminal Anthropology* (1895), "the results of a study of 6,034 living criminals," he summarized his findings:

"In Assassins we have prominent jaws, widely separated cheekbones, thick dark hair, scanty beard, and a pallid face.

"Assailants have brachycephaly [a rounded skull] and long hands; narrow foreheads are rare among them.

"Rapists have short hands…and narrow foreheads. There is a predominance of light hair, with abnormalities of the genital organs and of the nose.

"In Highwaymen, as in Thieves, anomalies of skull measurement and thick hair; scanty beards are rare.

"Arsonists have long extremities, a small head, and less than normal weight.

"Swindlers are distinguished by their large jaws and prominent cheekbones; they are heavy in weight, with pale, immobile faces.

"Pickpockets have long hands; they are tall, with black hair and scanty beards."

Lombroso's first book came under bitter attack from those who accused him – justifiably – of over-simplification. At the same time, some support for his theory of hereditary criminal types came the following year, with the publication of *The Jukes* by American sociologist Richard Dugdale. The founder of the Jukes line, a highly disreputable character, was born in New York early in the 18th century, and Dugdale claimed to have traced 700 of his descendants, all but a few of whom had become criminals or prostitutes.

In the opposing camp, a particularly strong critic of Lombroso was the

Cesare Lombroso, the Italian professor of forensic medicine. His first book, L'Uomo Delinquente, *was published in 1876 and introduced the theory that different types of criminals could be detected by their physical characteristics.*

Frenchman Alexandre Lacassagne, professor of forensic medicine at Lyon, who maintained that the causes of crime were social and declared, "every society has the criminals it deserves." Lombroso subsequently modified his theories, and in *Crime: Its Causes and Remedies* (1899) he pointed out findings that partly supported Lacassagne's suggestion: when food is readily available, crimes against property decrease, while crimes against the person, particularly rape, increase. Indeed, toward the end of his life, Lombroso recognized that the "criminal type" could no longer be distinguished simply by physical characteristics alone.

ANTHROPOMETRY

Lombroso's original theories were a development of anthropometry, a branch of anthropology that arose following the publication of Charles Darwin's *The Origin of Species* in 1859. Anthropometry's devotees spent their time taking

The German pathologist Rudolf Virchow in his laboratory at the Berlin Pathological Institute. It was his observation of unusual features in the skulls of criminals that inspired Lombroso to study the appearance of more than 6,000 living criminals over a period of more than 20 years.

The phrenological studies of the Austrian physician Franz Joseph Gall attracted much popular attention in the early 19th century. Enthusiastic supporters of his theories gave lectures and demonstrations, in crowded rooms, to people of all ages and professions.

physical measurements of human beings, and particularly their skeletons, in the hope of supporting – or refuting – Darwin's theories about the evolution of humankind. One of those who applied the principles of anthropometry to the practice of criminal investigation was Alphonse Bertillon.

Anthropometry at first attracted the attention of other criminologists, but it soon fell into disuse, when fingerprinting was internationally accepted as the sure method of identifying criminals. However, fingerprint analysis, like the Bertillon set of physical measurements, serves only as a means of identifying a previously

convicted person, as well as being the means of connecting a suspect with the scene of a crime. Because it does not provide a way of detecting the possibility that a person may be genetically disposed to commit crime, some experts have continued to search for a connection between visible physical characteristics and the criminal personality. (In this respect, it must be pointed out that practitioners of palmistry – a subject that is regarded as little better than witchcraft by the police and criminologists – claim to be able to detect psychological tendencies in the pattern of lines in the human hand.)

ALPHONSE BERTILLON

At the time of the publication of Lombroso's first book, the president of the Paris Anthropological Society was Dr. Louis Adolphe Bertillon, who devoted his studies to comparing and classifying the shape and size of the skulls of different racial types. His son Alphonse (1853–1914) at first showed little interest in his father's work. When he was appointed a junior clerk in the records office of the Police Prefecture, however, he realized that anthropological methods could be used to link newly-arrested people to previous crimes. One of his father's associates, the Belgian statistician Lambert Quetelet, had stated that no two people shared exactly the same combination of physical measurements, and young Bertillon proposed a related system of identification to his superiors.

Between November 1882 and February 1883, Bertillon painstakingly assembled a file-card system of 1,600 records, cross-referencing them with measurements he made on arrested criminals. It was on February 20, 1883, that he had his first success. A man calling himself "Dupont" was brought to him and, after taking his physical measurements, Bertillon began to go through his files. At last, triumphantly, he picked out a single card: "You were arrested on December 15th last year!" he exclaimed. "At that time you called yourself Martin." The news of this success made headlines in the Paris newspapers. By the end of the year, Bertillon had identified some 50 recidivists, and in 1884 he identified more than 300. Police and prison authorities throughout France swiftly adopted "Bertillonage."

Bertillon then began to make use of photography, both of arrested suspects and of the scenes of crimes. He established the procedure of taking portraits both full face and in profile – still the standard practice today – and also introduced what he called the *portrait parlé* (the "speaking likeness"). This was a system of describing the

Photography was a relatively new technique that was eagerly adopted by the young Alphonse Bertillon (above). It became a valuable adjunct to his *Bertillonage* system and his *portrait parlé*.

shape of facial features such as the nose, eyes, mouth, and jaw, and remains the basis of Identikit and other more modern identification systems.

Bertillon was at one time credited with the adoption of fingerprinting techniques, but in fact, although he sometimes recorded criminals' fingerprints, he remained convinced of the superiority of his measurement system, and on more than one occasion missed the identity of prints on file. As other countries took up fingerprinting in the early years of the 20th century, the French system of Bertillonage was eventually discarded.

PHYSIQUE AND CHARACTER

The German psychiatrist Ernst Kretschmer published *Physique and Character* in the early years of the 20th century. In his book he described his researches in this subject, but it was not until as late as 1949 that the American William Sheldon, in *Varieties of Delinquent Youth,* made the first systematic linking of body types with delinquency. He claimed that all people were of one of three basic types:

Endomorphs: generally soft, rounded, and plump, and characterized as friendly and sociable and loving comfort.

Mesomorphs: hard, muscular, and athletic, with a strongly developed skeleton. The personality is strong and assertive, with a tendency to be aggressive and, occasionally, explosive.

Ectomorphs: thin, weak, and generally somewhat frail, with a small skeleton and weak muscles. They tend to be hypersensitive, shy, cold, and unsociable.

Sheldon examined 200 men in a rehabilitation unit in Boston, compared them with a study of 4,000 students, and came to the conclusion that delinquents tended to be mesomorphs. This theory was further examined in *Unraveling Juvenile Delinquency* (1950) by Eleanor and Sheldon Glueck, who at first found some support for it, but eventually concluded that delinquency was related to a wider combination of biological, environmental, and pyschological factors.

ALIENATED MAN

Early in the 20th century criminologists began to turn their attention away from the physical characteristics of the "criminal type" and toward the mental processes – the psychology – that led people to crime. They were encouraged by the new ideas being put forward by the Vienna school of psychologists, led by Sigmund Freud and Carl Jung (see Chapter 2).

The English sociologist Dr. Charles Goring was one of the principal critics of Lombroso's theories. Dr. Goring reported that he had found as many cases of Lombroso's physical types among English

Here, a police officer takes Bertillonage measurements of a suspect's ear at New York Police Department headquarters in 1908.

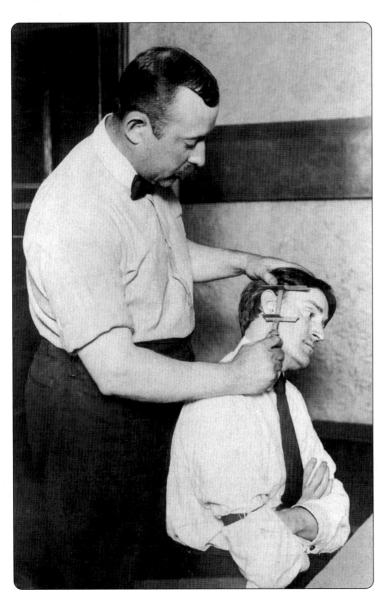

Bertillon's interest in photography led him to develop his "ladder camera," raised sufficiently high to enable him to photograph the whole body of a dead person as it lay where it had fallen. Subsequently, photographs of crime scenes have become ever more important in investigations, and they are studied with care by modern psychological profilers.

The modus operandi (MO) of a criminal – the type of tool used in a break-in, the way in which a murder is carried out, and many other characteristic factors – can be a valuable indication that can lead to the identification of the perpetrator.

his society, misunderstood, and therefore believes himself justified in following his own rules of behavior and conduct. The concept of alienation has become a vital part in the psychological assessment of criminality.

THE CRIMINAL'S METHOD

Since the mid-19th century, police investigators have realized that the handiwork of many persistent criminals can be recognized from what is generally known as their modus operandi ("method of working," usually abbreviated MO). The way in which a building is entered; the way in which a safe is broken open; the tools used; the type of explosive employed – or, in the case of murder, the way in which the victim is captured, killed, and perhaps mutilated – all these can provide clear indications that a succession of crimes have been committed by the same hand.

In cases of serial murder, the killer often leaves a characteristic "signature" – the way in which the body is disposed of, or some other unusual evidence – at the scene of the crime.

This signature should not be confused, however, with the MO. The MO is learned behavior, becoming modified and perfected as the offender becomes more experienced. The signature, on the other hand, is something that the criminal has to do to reach emotional fulfillment. It is not absolutely necessary for the successful accomplishment of the crime, but is part of the reason why he undertakes the crime in the first place.

university students as among convicts. Developing his argument in *The English Convict* (1913), Goring argued that many criminals were of inferior intelligence, and made the direct connection between this and crime.

This is as sweeping a generalization as Lombroso's classification of "atavistic" types, and there are many cases in which it is obviously not true. But Goring also identified what he called the "lone wolf" – or what the economist and philosopher Karl Marx had named the "alienated man." This is someone who feels isolated from

CASE STUDY: JACK THE RIPPER

Even after more than a century, "Jack the Ripper" continues to fascinate professional profilers. The crimes committed by this Victorian murderer provoked a wave of panic in the East End of London in the second half of 1888. The case is of particular interest to criminal profilers because it resulted in the first documented attempt at a psychological profile of a serial killer.

Between August and November, five women – all known prostitutes – had their throats brutally slashed, after which the bodies of four were then horribly mutilated. On the morning after the first killing, a newspaper report stated, "No murder was ever more ferociously and more brutally done." In this killing, and another that followed within a week, the woman's abdomen was ripped open, but the mutilations were soon to become even more terrible. Popular fear was heightened three weeks later, when the Central News Agency received a letter with the signature "Jack the Ripper," written in red ink.

IT'S IN THE BLOOD

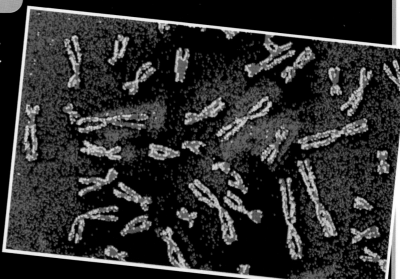

As researchers continued their remarkable advances in the science of genetics during the 20th century, they made what at first seemed to be an exciting discovery. In human beings, all the genetic information is held on 23 pairs of chromosomes, which control such physical factors as the color of hair and eyes, the structure of the body, and so on. One pair of chromosomes determines the sex of the individual: in the normal female, these are denominated XX, and in the normal male, XY.

In the male, the X chromosome comes from his mother, and the Y chromosome from his father. However, some males are found to have a combination of three chromosomes, either XXY or XYY. As the Y chromosome was linked with masculinity, it was suggested that an XYY male would be a "supermale," likely to be more aggressive, and possibly criminal. A report published in 1965 stated that there was a higher proportion of XYY chromosome in men confined in mental institutions than among the general population, and claimed that they had "dangerous, violent, or criminal propensities."

However, later studies showed that although a high proportion of XYY men had committed crimes, these were mostly petty property offenses, and they were no more likely to commit violent crimes than normal XY men.

This colored thermograph shows pairs of chromosomes.

No. 1423.—Vol. 55

THE PENNY
ILLUSTRATED PAPER
AND ILLUSTRATED TIMES

SEPTEMBER 8, 1888

REGISTERED AT THE GENERAL POST-OFFICE AS A NEWSPAPER.

London : Printed and Published at the Office, 10, Milford-lane, Strand, in the Parish of St. Clement Danes, in the County of Middlesex, by THOMAS FOX, 10, Milford-lane, Strand, aforesaid

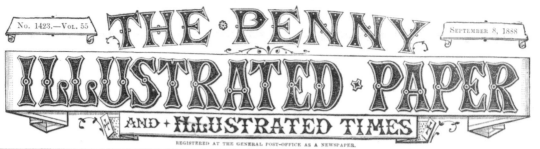

P.C. NIEL. J. 97.　DR LLEWELLYN　INSPR HELSON　THE CORONER

SKETCHES AT THE INQUEST

EAST London has a terror that must be stamped out. We illustrate on this page, and describe in another, Police-Constable Niel's discovery of murdered Mary Ann Nicholls in Buck's-row, Whitechapel, on the early morning of August the Thirty-first. This crime has so many points of similarity with the murders of the two other women in the same neighbourhood—one, Martha Turner, as recently as Aug. 7, and the other less than twelve months previously—that the police admit their belief that the three crimes are the work of one individual. All three women were of the same class, and each of them was so poor that robbery could have formed no motive for the crime. The three murders were committed within a distance of 200 yards of each other

THE WHITECHAPEL MYSTERY.

The letter read: "You will soon hear of me with my funny little games. I saved some of the proper red stuff in a ginger beer bottle over the last job to write with but it went thick like glue and I can't use it. Red ink is fit enough I hope, ha ha."

On the last day of September, just two days after this letter was received, the bodies of two more women were discovered. The next day, the Central News Agency received a postcard, in the same handwriting as the letter, and apparently bloodstained. It claimed:

"you'll hear about saucy Jackys work tomorrow double event this time…."

Most experts are now convinced that the letter and postcard were a hoax, perpetrated by a journalist to heighten interest in the case, but another letter, apparently in a different handwriting, was sent two weeks later to a member of a hastily created Vigilance Committee. Dated "From Hell," and signed, "Catch me when you can," it contained a horrific trophy – half a human kidney.

The fifth killing – the last positively attributed to the Ripper – was the most gruesome of all. The murder took place in the woman's rented room, and the Ripper had plenty of time to carry out his bloody work. The head was almost completely severed, parts of the body were cut off, and much of the flesh was stripped away and placed on a table nearby in a welter of blood. By this time, the police were speculating whether the murderer might even be a member of the medical profession.

The police surgeon who assisted in the autopsy on the fifth victim was Dr. Thomas Bond. He was originally called in to give an opinion on the Ripper's knowledge of surgery, but he went on to provide the police with a description of the killer. Affirming that all five murders were committed by the same person, he told police investigators: "The murderer must have been a man of

Far left: Sensational popular newspapers, sold for a penny, featured imaginative engravings of the murders committed by Jack the Ripper in the Whitechapel area of East London. This drawing is the artist's impression of the discovery of the body of the first victim, on August 31, 1888, by a patrolling policeman.

Left: This photograph of a passage and stairway shows the location of Jack the Ripper's fifth murder.

physical strength, and great coolness and daring. There is no evidence that he had an accomplice. He must, in my opinion, be a man subject to periodic attacks of homicidal and erotic mania. The character of the mutilations indicate that the man may be in a condition sexually, that may be called Satyriasis. It is of course possible that the Homicidal impulse may have developed from a revengeful or brooding condition of mind, or that religious mania may have been the original disease, but I do not think that either hypothesis is likely.

"The murderer in external appearance is quite likely to be a quiet inoffensive looking man, probably middle-aged, and neatly and respectably dressed. I think he might be in the habit of wearing a cloak or overcoat, or he could hardly have escaped notice in the streets if the blood on his hands or clothes were visible.

"Assuming the murderer to be such a person as I have just described, he would be solitary and eccentric in his habits, also he is likely to be a man without regular occupation, but with some small income or pension. He is possibly living among respectable persons who have some knowledge of his character and habits and who may have grounds for suspicion that he is not quite right in his mind at times. Such persons would probably be unwilling to communicate suspicions to the Police for fear of trouble or notoriety, whereas if there were prospect of reward it might overcome their scruples."

Considering the possibility that the Ripper might have – at least – occupational

> "One of the requisites necessary to enable an investigating officer to work with accuracy is a profound knowledge of men."

experience of cutting and gutting meat or fish, Dr. Bond argued: "In each case, the mutilation was inflicted by a person who had no scientific nor anatomical knowledge. In my opinion, he does not even possess the technical knowledge of a butcher or horse slaughterer, or any person accustomed to cutting up dead animals."

In spite of the guidance provided by Dr. Bond, the police never apprehended the Ripper. There were other murders, but none bore his distinctive handiwork. Over the years, dozens of possible suspects have been named by researchers, from a mad midwife to Prince Albert Victor, the grandson of Queen Victoria. Crime writer Patricia Cornwell, in her book *Jack the Ripper: Portrait of a Killer* (2002), has stated her conviction that the famous painter Walter Sickert was the murderer.

Even the FBI offered an assessment on the occasion of the "Ripper" centenary – one that was strikingly similar to Dr. Bond's (see Chapter 3). Only one thing is certain: the identity of Jack the Ripper will never be proven.

DR. HANS GROSS

The formal investigation of crime was first detailed by Dr. Hans Gross in his *System der Kriminalistik* (1893). This was originally published in English as *Criminal Investigation* in 1907 and it remained the bible of police investigators for more than half a century. Gross (1847–1915) was an Austrian magistrate who stressed that crime investigation is a science and a technology. He became professor of criminology at the University of Prague, and then professor of penal law at the University of Graz.

Gross wrote: "One of the requisites necessary to enable an investigating officer to work with accuracy is a profound knowledge of men.... To this end, everything in life can be utilized: every conversation, every concise statement, every word thrown out by chance, every action, every aspiration, every trait of character, every item of conduct, every look or gesture, observed in others...." He suggested that the experienced investigator might well be able to infer the personality characteristics of a perpetrator from the manner and nature of the crime, that is, from his MO.

By the end of the 19th century, the foundations of the psychological assessment of criminals had been laid.

By the end of September 1888, four prostitutes had died at the hands of the vicious killer Jack the Ripper. These four murders were committed in the open, as in this illustration, but the fifth – and the last to be attributed to "Jack" – took place in the woman's lodging.

CONAN DOYLE, SHERLOCK HOLMES, AND DR. JOSEPH BELL

For well over a century Sherlock Holmes has remained the most famous detective in fiction. The creation of Sir Arthur Conan Doyle, he appeared in four novels and 56 short stories published over a period of 40 years, from 1887 to 1927. Much of Holmes's brilliant detective work was attributed to his technical expertise. Doyle, through the conversations that took place between Holmes and Dr. Watson, delighted in explaining exactly how the great detective employed his analytical powers to deduce a person's nature and profession, or a criminal's personality.

In developing a character for his first story, Doyle wrote:

Dr. Joseph Bell, Conan Doyle's mentor at the Edinburgh Royal Infirmary, whose "eerie trick of spotting details" was taken as the model for Sherlock Holmes.

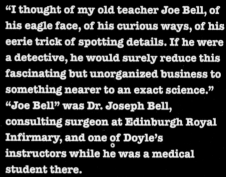

"I thought of my old teacher Joe Bell, of his eagle face, of his curious ways, of his eerie trick of spotting details. If he were a detective, he would surely reduce this fascinating but unorganized business to something nearer to an exact science." "Joe Bell" was Dr. Joseph Bell, consulting surgeon at Edinburgh Royal Infirmary, and one of Doyle's instructors while he was a medical student there.

"For some reason which I have never understood he singled me out from the drove of students who frequented his wards and made me his out-patient clerk, which meant that I had to array his out-patients, make simple notes of their cases, and then show them in, one by one, to the large room in which Bell sat.... Then I had ample chance of studying his methods and of noticing that he often learned more of the patient by a few quick glances than I had done by my questions. Occasionally the results were very dramatic.... In one of his best cases he said to a civilian patient: 'Well, my man, you've served in the army.'

'Aye, sir.'
'Not long discharged?'
'No, sir.'
'A Highland regiment?'
'Aye, sir.'
'A non.com officer?'
'Aye, sir.'
'Stationed at Barbados?'
'Aye, sir.'
'You see, gentlemen,' he would explain, 'the man was a respectful man but did not remove his hat. They do not

in the army, but he would have learned civilian ways had he been long discharged. He has an air of authority and he is obviously Scottish. As to Barbados, his complaint is elephantiasis, which is West Indian and not British.' To his audience of Watsons it all seemed very miraculous until it was explained."

A few years after the first appearance of Sherlock Holmes in print, Dr. Bell himself confirmed his methods in a letter to a representative of the *Strand* magazine, in which many of Doyle's stories were published. He wrote:

"Physiognomy helps you to detect nationality, accent to district, and, to an educated ear, almost to county. Nearly every handicraft writes its sign manual on the hands. The scars of the miner differ from those of the quarryman. The carpenter's callosities are not those of the mason. The shoemaker and the tailor are quite different. The soldier and the sailor differ in gait, though last month I had to tell a man who said he was a soldier that he had been a sailor in his boyhood. The subject is endless: the tattoo marks on hand or arm will tell their own tale as to voyages; the ornaments on the watch chain of the successful settler will tell you where he made his money...."

It was from his early observation of Dr. Bell at work that Doyle originally derived Sherlock Holmes's deductive methods, but – with a mixture of ingenuity and imaginative analysis – he developed the technique to a point where it could serve as an example to a real-life investigator. The great French criminologist Dr. Edmond Locard, one of the founders of modern forensic science, was a fan of Sherlock Holmes, and recommended a study of Doyle's stories.

Sherlock Holmes was introduced to the reading public in *A Study in Scarlet*, in which he first demonstrated his formidable investigative powers.

THE PSYCHOLOGISTS INVESTIGATE

The famous Swiss psychologist Carl Gustav Jung (left) came to the conclusion that there was a "collective unconscious" that reflected the atavistic instincts of primitive humankind and divided personality into more than a dozen "archetypes."

In spite of many sociological studies of criminality that were made during the first quarter of the 20th century, the first reported face-to-face psychological interview with a convicted killer was not made until 1930. This was conducted by the German psychiatrist Professor Karl Berg, who questioned serial murderer Peter Kürten, the "Vampire of Düsseldorf," in prison before his execution.

Kürten, born in 1883, was described by his wife (whom he married in 1923) and by his neighbors as a mild, conservative, soft-spoken man, a churchgoer, and a lover of children. At his trial, he wore a neat suit and, it is said, smelled of eau de Cologne. Yet Kürten was charged with nine counts of murder and seven other assaults with intent to kill, and he had confessed during his police interrogation to an orgy of rape, mutilation, and murder.

His criminal career began at age 17, when he was sentenced to two years in jail for petty theft. Many other sentences followed – in fact, Kürten spent 20 of his 47 years in prison. It was in 1913, when he was released from jail following seven years of hard labor, that he committed his first murder. He found a 13-year-old girl lying asleep near the window of her father's inn at Köln-Mulheim, and he strangled her before cutting her throat with a pocketknife. The child's uncle was

In 1929, a wave of horrific murders caused alarm throughout Düsseldorf. Here, a group of citizens reads a police notice about the latest killing.

charged with the crime, but the evidence was considered insufficient to convict him – and Kürten went unsuspected.

Very little is known of Kürten's activities in the years that followed; he is credited with a series of sexual assaults upon women, although few were reported to the authorities, and Kürten was not identified. Then, in 1929, a wave of near-hysteria swept through Düsseldorf as a succession of 23 attacks occurred, mostly on young women and girls, In February of that year, a young woman was stabbed 24 times with a pair of scissors, but her screams scared off her assailant, and she survived. Six days later, a nine-year-old girl was stabbed to death, also with a pair of scissors, and an attempt made to incinerate her body with kerosene. After another four days, a male drunk was killed in the same manner, and his killer drank the blood that spurted from his wounds.

In August, a young woman suffered a similar fate. This was followed by two unsuccessful attempts at murder, where the intended victims were fortunately able to escape and give a partial description of their assailant to the police. However, more killings followed, including two five-year-old girls. The killer had now abandoned his scissors, and used a hammer to batter his victims to death.

VAMPIRE'S MISTAKE

In May 1930, the "Vampire of Düsseldorf" made a fatal mistake. He picked up a young woman from Cologne, 20-year-old Maria Budlies, who had just arrived in the city looking for work. He took her to the apartment he shared with his wife, where he attempted to rape her, and then to a nearby wood, where he began to strangle her. Suddenly, he released his grip and asked her if she remembered his

Photographs of Peter Kürten, taken by the police after his arrest, reveal him as an unexpectedly calm and self-confident individual – the very model of a responsible citizen.

address. When she had assured him that she did not, he unaccountably let her go.

Maria did not report the incident to the police, but described it in a letter to a friend. By a happy coincidence, the letter had been incorrectly addressed, and was opened by a clerk at the post office. He immediately handed it to the police, who tracked down Maria – and she, in fact, remembered significant details of the man's address. Taken there by detectives, she was shocked to see her attacker leaving the premises – a man, the landlady said, named Peter Kürten. The next day, Kürten's wife went to the police and told them that her husband was the feared Vampire of Düsseldorf.

At his trial Kürten pleaded insanity, but the plea was dismissed; he was found guilty on all counts, and sentenced to be beheaded. In prison, he talked frankly with Professor Karl Berg. He said he was one of 13 children born to a sand-molder and his wife in Köln-Mulheim. Kürten's father was a drunkard who beat and abused both his wife and his children, and later was sentenced to jail for attempted incest. Even as a young child, Kürten showed a violent disposition, attempting to drown one of his playmates by holding his head under water.

Professor Berg reported that Kürten was a very early sexual developer. He said that he had begun to derive sexual pleasure from assisting the local dogcatcher in torturing and killing animals. At the age of 13 he started practicing bestiality. Later, he became an arsonist: "The sight of the flames delighted me," he said, "but above all it was the excitement of the attempts to extinguish the fire, and the agitation of those who saw their property being destroyed."

The psychiatrist was struck by the killer's frankness and intelligence and his ability to recall his activities over the past 20 years. At the age of 16, said Kürten, he had visited the Chamber of Horrors in a local waxworks museum. "I am going to be someone famous like those men, one of these days," he told a friend. Later, the story of Jack the Ripper excited him. "When I came to think over what I had read, when I was in prison," he said, "I thought what pleasure it would give me to do things of that kind when I got out again." He told Berg that he had stared with longing at the bare throat of the stenographer who recorded his confession, and was filled with the desire to strangle her.

At his trial, Kürten stated: "I did not kill either people I hated or people I loved. I killed whoever crossed my path at the moment my urge for murder took hold of me."

He ate a hearty meal before he was beheaded, and it is said that he turned to his executioner on the scaffold and asked: "Will I still be able to hear for a moment the sound of my blood gushing?

That would be the pleasure to end all pleasures." Professor Berg described him as a "narcissistic psychopath," and "a king of sexual perverts."

FREUD, JUNG, AND ADLER

Karl Berg wrote a landmark book on the case of Peter Kürten, but it was not published in English (as *The Sadist*) until 1945. Meanwhile, psychologists devoted their attention to adapting mainstream theories to the study of criminality.

During the 1930s and 1940s, the principal theories in psychology were those of Sigmund Freud (1856–1939), his former collaborator and colleague Carl Gustav Jung (1875–1961), and Alfred Adler (1870–1937). Freud was the pioneer among this group – all three of whom were psychiatric practitioners in Vienna – and he began to develop his theories by referring to his own experiences and emotions. As he recalled incidents from his childhood and compared them with those of his patients, he came to the conclusion that sexual instincts were present in infants from the very beginning, and could develop in different ways depending upon influences within the family. This resulted in his concept of the "Oedipus complex," his description of the erotic feelings of a son for his mother and concomitant sense of competition with his father.

Freud began to analyze his dreams and those related to him by his patients. In his major work, *Die Traumdeutung* (*The Interpretation of Dreams*, 1900), he stated that dreams are disguised wish fulfillment – the evidence of repressed sexual desires. Despite the hostility of the medical establishment he persisted in this view, finding manifestations of sexual instincts in every instance of social, and antisocial, behavior. In later publications he developed theories of the division of the unconscious mind into three parts: the "id," which is associated with instinctive impulses and the desire for immediate satisfaction of primitive needs; the "ego," which is aware of itself, is most in touch with external reality, and exerts control over behavior; and the "super-ego," which is a largely unconscious element that incorporates the absorbed moral standards of the family and wider community, and includes the conscience.

Freud and his family escaped from the

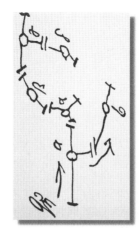

clutches of the Gestapo when the Nazis overran Austria in 1938, and settled in England, where his daughter Anna became an eminent child psychoanalyst.

Jung, a younger man, originally from Switzerland, was for some years a disciple of Freud, but in 1913 he broke away from Freudian theories and developed ideas that, ironically, reflected an approach to psychology that was reminiscent of Lombroso's discredited anthropometry. Analyzing his own dreams, Jung came to the conclusion that there was a "collective unconscious" that reflected the "atavistic" instincts of primitive

Far left: At his trial, Peter Kürten appeared in a neatly buttoned suit, with a meticulously knotted necktie. It was reported that he had even sprayed himself with eau de Cologne.

Left: A diagram sketched by Freud to indicate the way in which he believed repression worked.

Sigmund Freud's pioneering work on the interpretation of dreams, and the stress he placed on their significance as repressed sexual desires, laid the foundations of modern psychological theory.

Carl Jung was a disciple of Freud for some years, but later broke away to establish his own theories. Jung's belief in the "collective unconscious," dating back to the primitive instincts of early humankind, led him to investigate such unexpected subjects as alchemy and UFOs.

humankind. He divided personality into over a dozen "archetypes." He believed that more than one of these types is likely to be revealed in any individual. Jung is also responsible for the distinction between "extroverts," who possess an aptitude for dealing with the external world and other people, and "introverts," who concentrate their feelings inward upon themselves.

Adler, too, began as a Freudian analyst, but in 1911 he also broke away and

founded his own group. He had been small and weak as a child. In 1907, he published a groundbreaking work, "Organ Inferiority and Its Psychical Compensation," in which he argued that people born with a physical defect – such as deafness, poor vision, or lameness – would unconsciously adjust the way in which their personalities were developing to compensate for this inferiority. In its way, this theory could also owe something to Lombroso. From this fundamental observation, Adler

ADLER ON CRIME

Alfred Adler wrote that "the criminal is not interested in others." Fundamentally, he believed that "all criminals are cowards," evading problems that they were unable to solve in a socially acceptable way. Crime, he said, was a coward's imitation of heroism. When criminals were found out, they believed that it was because they had not been quite clever enough, or because of sheer bad luck; "but next time I shall outwit them." And if they escaped apprehension, their feeling of superiority was strengthened by the admiration of their associates. Adler stated that even the death penalty was no deterrent, because the criminal remained convinced that he would not be caught, due to his superior shrewdness.

developed the concept of the "inferiority complex," a human drive to become superior in compensation for feelings of inferiority. This might or might not derive from any physical failing, for he believed that all children – even those brought up in well-balanced family environments – could suffer from such feelings. He maintained that the "will to power" was normal, and more important in determining behavior than the primitive sexual drive.

Beginning in 1920, Adler held child guidance clinics, in which children, their parents, and teachers were counseled. In 1932 he joined the faculty of the then Long Island College of Medicine.

Another influential psychologist was the German Richard von Krafft-Ebing (1840–1902). He carried out a wide-reaching study of sexual behavior, resulting in his publication of *Psychopathia Sexualis* (*The Psychopathology of Sexuality*) in 1886. Krafft-Ebing identified the perverted desire to inflict pain or humiliation upon the object of sexual desire, and its reverse – deriving sexual pleasure from personal suffering – and applied the names of "sadism" and "masochism" to its respectively active and passive forms.

These, then, were the main psychological principles upon which criminologists carried out their studies during the 1930s and 1940s. Despite numerous divergent theories that have been advanced in recent times, they remain the principal foundation of modern criminal psychology.

> Adler held that the "will to power" was normal, and more important than the primitive sexual drive central to Freudian theory.

The studies of psychopathic sexuality by Richard von Krafft-Ebing predated those of Sigmund Freud by more than 10 years. In particular, he was responsible for the definition of the concepts of sadism and masochism.

THE CASE OF THE MAD BOMBER

In November 1940, a small unexploded homemade pipe bomb was found on a windowsill inside the premises of Consolidated Edison, the company that supplied electricity for New York City. With it was a neatly lettered note: "CON EDISON CROOKS, THIS IS FOR YOU!" At that time the police did not consider the matter to be very serious, but they changed their minds when a similar bomb – fitted with an alarm clock detonator that had not been wound up – was found in the street 10 months later.

In December 1941, the Japanese attacked Pearl Harbor, and shortly afterward the police received a letter, mailed from Westchester County, outside New York City. Written once again in neat capitals, it read: "I WILL MAKE NO MORE BOMB UNITS FOR THE DURATION. OF THE WAR – MY PATRIOTIC FEELINGS HAVE MADE ME DECIDE THIS – I WILL BRING CON EDISON TO JUSTICE LATER – THEY WILL PAY FOR THEIR DASTARDLY DEEDS. F. P."

Over the next five years there were no further bombs, although a series of 16 similar notes were received for a time by Con Ed, newspapers, department stores, and hotels. The police began to suspect that "F. P." had abandoned his campaign, or perhaps had died. Then, on March 25, 1950, another unexploded bomb was found in New York's Grand Central Station.

All three bombs had been carefully crafted, and police officials hoped that the "Mad Bomber," as he quickly became known, did not intend them to detonate. However, the next, planted in a telephone booth at the New York Public Library, did explode – fortunately without injuring anybody. Letters received by newspapers threatened more: "TO SERVE JUSTICE."

From 1951 through 1954, 12 more bombs exploded – at Radio City Music Hall, the Port Authority bus terminal, Rockefeller Center, and many other locations. It was the last bomb of 1954, hidden in a seat in a movie theater, that produced the first injuries, when four people were slightly hurt.

Six bombs were planted in 1955, two of which failed to explode. The bombs were becoming increasingly destructive, and the Mad Bomber was obviously becoming angrier. He sent more letters to newspapers and even made telephone calls, but his voice was quiet, anonymous, and unidentifiable. In a letter to the *Herald Tribune* he concluded (as always, in capital letters): "SO FAR 54 BOMBS PLANTED – 4 TELEPHONE CALLS MADE – THESE BOMBINGS WILL CONTINUE UNTIL CON EDISON IS BROUGHT TO JUSTICE."

Then, on December 2, 1956, a bomb exploded in the Paramount Theater in Brooklyn, injuring six people, three of them seriously; doctors worked all night to save the life of one. Just over three weeks later, the editor of the *Journal-American*

published an open letter to the Mad Bomber, begging him to give himself up, and offering full publicity for his grievances. After two days, a reply was received; "start working" on three retired administrators, it suggested, including the former governor. The letter also listed 14 bombs set during 1956, many of which had not been discovered.

Soon, another letter arrived, which provided some clues as to the identity of the bomber:

As the activities of the Mad Bomber accelerated from 1951 through 1954, 12 of his bombs exploded in locations throughout New York City, while others remained undiscovered. One bomb was detonated in Radio City Music Hall – fortunately, without anybody being injured.

"I was injured on a job at Consolidated Edison Plant – as a result I am adjudged totally and permanently disabled. I did not receive any aid of any kind from the company – that I did not pay for myself – while fighting for my life…."

Because the possibility of finding helpful company records seemed slight – the bomber had been intermittently active for 16 years – Inspector Howard E. Finney of the New York Police Crime Laboratory took the revolutionary decision to consult a psychiatrist. The man he chose was Dr. James A. Brussel, who had many years' experience with the criminally insane.

Brussel first deduced that the writer of the letters was not a native American, for the letters contained no informal Americanisms; and phrases such as "they will pay for their dastardly deeds" indicated that he was of an older generation. The style of the letters was not German, or Latin, which eventually led Brussel to suggest that they had been written by someone of Slavic origin, either an immigrant or the son of an immigrant.

The bomber, said Brussel, was a paranoiac, a person with a psychosis characterized by delusions of persecution. Since Brussel believed that paranoia developed in people in their thirties (not a generally accepted view), he calculated that the man was now around 50. Like all paranoiacs, even though he was likely to be on the verge of insanity, he was careful, meticulous, and highly controlled. The neat capitals of the letters supported this supposition, but Brussel was struck by the unusual shape of the Ws, which were made up of two Us. A practitioner of Freudian psychology, he suggested that these looked like female breasts, and deduced that the man had a fixation on his mother.

Brussel's summary read: "Single man between 40 and 50 years old, introvert. Unsocial but not antisocial. Skilled mechanic. Cunning. Neat with tools. Egotistical of mechanical skill. Contemptuous of other people. Resentful of criticism of his work but probably conceals resentment. Moral. Honest. Not interested in women. High school graduate. Expert in civil or military ordnance. Religious. Might flare up violently at work when criticized. Possible motive: discharge or reprimand. Feels superior to critics. Resentment keeps growing. Present or former Consolidated Edison worker. Probably case of progressive paranoia."

Much of this was already deducible from the letters, and some parts of the assessment were later found to be inaccurate, but other conclusions proved startlingly true. Brussel told the police that the wanted man was well built and clean-shaven. He was unmarried, probably living with an older female relative, perhaps his mother, and wore double-breasted suits, buttoned up.

Even as the police were considering these details, the *Journal-American* received a typed letter. It read, in part: "I was injured on the job at the Con Edison plant – September 5th, 1931…." There followed an intensive search of the

> The Mad Bomber, said Dr. Brussel, was a paranoiac, a person with a psychosis marked by delusions of persecution.

Left: Dr. James Brussel's profile of the "Mad Bomber" described him as well-built, clean-shaven, and around 50 years old. When George Metesky was finally arrested in January 1957, he was 52 years old and closely resembled Dr. Brussel's description.

After his arrest, George Metesky (center) cheerfully confessed that he was the Mad Bomber. He was found unfit to stand trial and was committed to a state hospital for the criminally insane.

surviving Consolidated Edison records, and by pure luck the relevant document was found. It named George Metesky, the son of a Polish immigrant born in 1904, who had been injured in a boiler explosion accident, which resulted in pneumonia and subsequent tuberculosis. He had received a paltry $180 in compensation. There was a large Polish community in Bridgeport, Connecticut – and Westchester County,

where the letters had been mailed, lay between Bridgeport and New York City.

In Waterbury, near Bridgeport, detectives armed with a warrant called at Metesky's home, where he lived with two elderly half-sisters. A mild-faced, sturdily built, middle-aged man wearing gold-rimmed glasses opened the door. As it was past midnight, he was wearing a robe over pajamas; the detectives told him to get

dressed, and when he reappeared he was wearing a shirt and tie and a neat double-breasted suit – buttoned up. A search of the garage revealed a workshop with a lathe and a length of the piping used to the make the bombs. In Metesky's bedroom officers found the typewriter he had used to write his last letter. At the police station, Metesky willingly confessed that he was the Mad Bomber. He was found unfit to stand trial, and was committed to a state hospital for the criminally insane, where he died soon after of tuberculosis. He told the police that the letters "F. P." stood for "Fair Play."

Dr. Brussel's success attracted attention from crime fighters, and he later published a description of his work, *Casebook of a Crime Psychiatrist* (1968). In this book, he described how, in 1964, he had been invited to join a panel of psychiatrists who were attempting to develop a useful psychological profile of the infamous "Boston Strangler."

THE BOSTON STRANGLER

Between June 1962 and January 1964, 13 women, mostly elderly, were murdered in the region of Boston, Massachusetts. In every case, they were sexually assaulted – sometimes bitten, bludgeoned, or stabbed – and their naked bodies laid out, as if for a pornographic photograph. Each was strangled – usually with an item of their own clothing, such as stockings or pantyhose – and the killer left his "signature" in the form of a neat bow tied beneath the victim's chin.

The majority opinion of the psychiatric panel appointed by the state authorities to assess the case was that two separate people, both unmarried, were responsible for the killings: one a schoolteacher and

the other a man living alone. Both were consumed with hatred of their mothers, who were probably deceased. The panel ventured that, during the childhood of the killers, their mothers "walked about half-exposed in their apartment, but punished them severely for any curiosity." These experiences led to the stranglers acting out their resentment in adulthood, murdering older women in a manner that was both "loving and sadistic." And both, the panel concluded, had a weak and distant father.

Dr. Brussel disagreed with these findings. His "psychofit" was of a strongly built 30-year-old, of average height, clean-shaven and dark-haired, probably of Spanish or Italian heritage. He detected gradual slight changes in the killer's MO: "Over the two-year period during which he has been committing these murders, he has gone through a series of upheavals – or, to put it another way, a single progressive upheaval. In this two-year period, he has

In 1968, Dr. James Brussel published his book, Casebook of a Crime Psychiatrist, *in which he recorded not only the case of George Metesky, but also his disagreement with the assessment of the panel of psychiatrists appointed to consider the case of the Boston Strangler. This book was an early inspiration for Agent Howard Teten, who introduced psychological analysis to the FBI.*

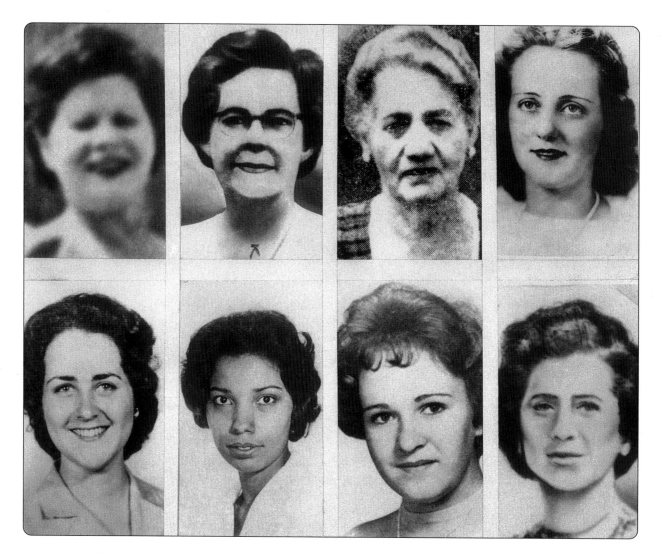

Some of the 13 female victims who were sexually assaulted and murdered in attacks attributed to the Boston Strangler between June 1962 and January 1964.

suddenly grown, psychosexually, from infancy to puberty to manhood."

The last killing, of 19-year-old Mary Sullivan, occurred on January 4, 1964. Nothing more was heard of the Strangler until October 27, when he entered the apartment of another woman, posing as a detective. He tied her to her bed, sexually assaulted her, and then uncharacteristically left, saying, "I'm sorry." Her description of the assailant matched that of a man already on police files.

A MAN IS CAUGHT

His name was Albert DeSalvo, age 33. Dark-haired and shaven, he was slightly below average height, but strongly built: he had been a U.S. Army welterweight champion while serving in Germany. There he married a girl named Irmgard from Frankfurt and was posted to Fort Dix, New Jersey. While still enlisted, he was charged with sexually molesting a 9-year-old girl. Her mother refused to prosecute, however, and the Army was unable to take the case further. On discharge, DeSalvo moved to Boston with his wife, where they raised two small children. Irmgard complained that he had an insatiable sex drive, and he later admitted that he thought about sex day and night.

Working as a local handyman, DeSalvo found many opportunities for breaking and entering, and in 1958 he was arrested and given a suspended sentence. Soon after, he began a sexual campaign that earned him the name of the "Measuring Man." Armed with a clipboard and tape measure, he would call on attractive young women in their apartments. He introduced himself as a representative of a modeling agency, which had selected them as possible models for television commercials. He did not attempt rape, but succeeded in

seducing many of them – sometimes, he claimed, even being seduced by them.

In March 1960 DeSalvo was arrested following a break-in and admitted that he was the Measuring Man. He received a two-year sentence and was released after 10 months. Police files listed him as a burglar, but not as a sexual deviant.

On his release he became more aggressive, breaking into apartments and tying up and raping his victims. Wearing a green shirt and work pants, he became known as the "Green Man." Extending his range throughout Massachusetts and Connecticut, he assaulted hundreds of women; he subsequently bragged that he had raped six in a single morning, and put the total at more than 1,000.

Following his identification from the

Albert DeSalvo, never officially identified as the Boston Strangler, admitted only to housebreaking and rape, and was committed to Boston State Hospital, diagnosed as a schizophrenic unfit to stand trial. He escaped from the hospital in 1967 and is pictured here after his recapture.

WAS DESALVO THE STRANGLER?

After hearing him talk constantly about sex and violence, an inmate in the Bridgewater mental institution came to the conclusion that DeSalvo was the Boston Strangler. He told his lawyer, the young F. Lee Bailey, who interviewed DeSalvo several times, tape-recording the conversations. The prisoner announced that he was indeed the Strangler, adding two more murders to the known count of 13. He also divulged details that had not been made public by the police – details that only the killer could have known.

Bailey was convinced, but the police had no eyewitnesses and little in the way of evidence needed to prosecute. Most experts concluded that DeSalvo was the Strangler – and one thing is certain: the killings ceased after his confinement.

Nevertheless, in December 2001, doubt was cast upon his identity as the Strangler. Mary Sullivan's body was exhumed for investigation by a team of forensic scientists at George Washington University. DNA analysis of stains from her underwear did not match her DNA, nor that of DeSalvo. According to Professor James Starrs, the head of the team: "The evidence that has been found is quite clearly indicative of the fact that Albert DeSalvo was not the rape-murderer of Mary Sullivan."

Miss Sullivan's nephew reported that he had personally confronted the man he believes murdered his aunt. The man was a suspect, but further investigation was dropped after DeSalvo's alleged confession. However, "Finding someone's DNA at a crime scene doesn't tell you how it got there, and when," said Frederick Bieber, associate professor of pathology at Harvard Medical School. "Similarly, not finding someone's DNA at a crime scene doesn't mean they weren't there."

F. Lee Bailey gained a reputation as an outstanding defense lawyer. Here, he holds up a photograph of DeSalvo.

description given by his last victim, DeSalvo was once more arrested by the police, but only on a charge for breaking and entering.

He was sent to the mental institution at Bridgewater, Massachusetts, although the police did not identify him as the Boston Strangler. He claimed to hear voices, and was diagnosed as schizophrenic. In February 1965 he was ordered to be detained indeterminately, and he was later transferred to Walpole State Prison, where he was found in his cell, stabbed through the heart, in November 1973.

The failure of the psychiatric panel to provide a detailed profile of the Boston Strangler set back the development of psychological profiling for several years. As John Godwin wrote in *Murder USA: The Ways We Kill Each Other* (1978):

"Nine out of ten of the profiles are vapid. They play at blind man's bluff, groping in all directions in the hope of touching a sleeve. Occasionally they do, but not firmly enough to seize it, for the behaviorists producing them must necessarily deal in generalities and types. But policemen can't arrest a type. They require hard data: names, dates, none of which the psychiatrists can offer."

In spite of these strong reservations, psychologists continued to be interested in the enormous potential of profiling, and in 1974 a British case provided them with convincing justification.

MURDER ON A SUNDAY MORNING

Early on the morning of Sunday, September 22, 1974, a young woman covered in blood staggered into the arms of a constable outside the police station at Chatham, in Kent, England. As she collapsed into unconsciousness, she whispered, "I was attacked by a man." She had clearly been stabbed in the stomach. A police dog-handler immediately retraced her steps with his dog, discovering a number of evidential traces on the way, to a footpath hidden by thick bushes – the point at which the girl had been assaulted. When she died shortly after admission to the hospital, her identity was quickly established as Susan Stevenson, who had been on her way to nearby Rochester Cathedral, where she was a bell-ringer and chorister.

Albert DeSalvo as he never appeared to his rape victims, in a neat suit, shirt, and necktie. In the years before his final arrest he had become known to police as the "Green Man," habitually dressed in a green shirt and work pants, and representing himself as a local handyman.

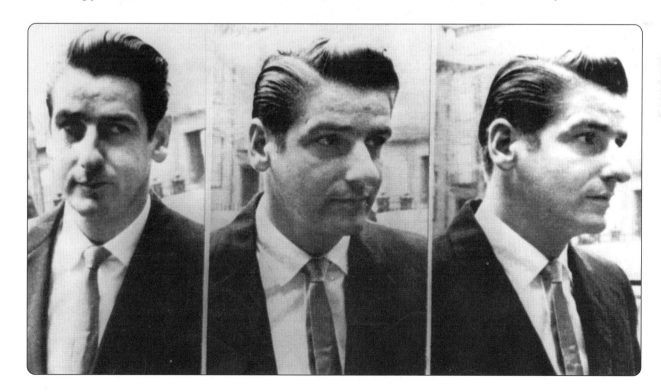

The murder inquiry was linked to two similar attacks in the previous year, but the police had little in the way of positive information. Professor J. M. Cameron, the leading pathologist who performed the autopsy, was able to describe the physical details of the attack, and consulted a colleague, the forensic psychiatrist Dr. Patrick Tooley, for more information. After visiting the scene of the crime and hearing the findings of Professor Cameron, Dr. Tooley wrote his assessment for the police:

"The man is aged between 20 and 35 years, possibly a psychopath with previous convictions. Generally one could expect too from his record that he had made a number of court appearances, was convicted at an early age, and had possibly been in a special home; likely to be a manual worker and either unemployed or frequently changing jobs. Previous convictions could include unlawful sexual intercourse, drunkenness, robbery, and assaults generally.

"Father absent – mother restrictive, sexually prudish, and devoted to son and spoils him. He, in turn, resents this and has a hate complex toward women. Despite that, he wants an affair with a woman but cannot make a normal approach. He does not mix socially and walks alone, in open spaces. He could be a 'peeper,' but seldom resorts to indecent exposure."

The police interviewed more than 6,000 people. After more than six weeks, the inquiry concentrated on a single man, a dock-laborer named Peter Stout. At an interview, he made repeated references to the Devil, blaming him for telling him to steal. He was arrested on suspicion of being the murderer, and, as journalist Tom Tulley described in his biography of Professor Cameron, *Clues to Murder* (1986):

"The background of Peter Stout was revealing. He was aged 19, single, and had one elder sister, two elder brothers, and a younger brother. Both his parents were dead. His father had been a drunkard and a bully, and was disliked by all the children, but they had all loved their mother. When Stout was 14 he was convicted of indecently assaulting a woman, and he himself had been the victim of attempted buggery when he was 10 years old. This fitted exactly into the pattern of the man described by Dr. Tooley, and there were other things that also fitted. He was a loner who went for long walks, and he did not mix well with others."

Stout's interrogation was long and drawn out, but at last he gave a statement that confirmed the sequence of the attack that had been detailed by Professor Cameron. The statement ended:

"All I can say is the Devil has got at me – and when he thinks I have my funny turns, he does some extra poking and things like that and I give in to him. I don't remember what I do when I have funny turns sometimes – they just come at any time."

At about the same time as the Peter Stout case, a major step forward in criminal profiling was being made in the United States by the FBI.

"All I can say is the Devil has got at me – and when he thinks I have my funny turns, he does some extra poking."

PROFILING HITLER

In 1943, during World War II, the U.S. Office of Strategic Services (the forerunner of the CIA) commissioned a "psycho-dynamic personality profile" of the German Führer, Adolf Hitler, from psychiatrist Walter Langer. They asked for "a realistic appraisal of the German situation. If Hitler is running the show, what kind of a person is he? What are his ambitions? We want to know about his psychological make-up – the things that make him tick. In addition, we ought to know what he might do if things begin to go against him." The intelligence authorities also needed the profile to be of help to them, should Hitler be captured, in developing an effective interrogation strategy.

Dr. Langer divided his report on Hitler into the following sections: "As he believes himself to be," "As the German people know him," "As his associates know him," "As he knows himself," "Psychological analysis and reconstruction," and "Hitler's probable behavior in the future."

After a detailed assessment of Hitler's personality – and in particular his psychological abnormalities – Langer provided a striking prediction of his likely fate. Death by natural causes was considered improbable, as, at that time, the Führer was believed to be in relatively good health. He might seek refuge in a neutral country (possibly in South America), but this also seemed unlikely, as he truly believed himself to be his country's figurehead and savior. After considering other possibilities – Hitler's assassination following a military coup or even death while personally leading his forces in a last desperate battle – Langer predicted that the Führer would commit suicide when defeat proved inevitable. And so it proved. In the last days of the war in May 1945, as Russian troops fought their way into Berlin and Allied forces advanced inexorably from the west, Hitler, holed up in his bunker beneath the German Chancellery, killed his companion, Eva Braun, and then himself.

Here, Adolph Hitler, the Führer (in German, "leader"), is amused by the presence of a small boy at a military parade in 1936.

A similar profiling approach became necessary after the end of the war, when a team was set up to identify wanted war criminals. Lionel Haward, a young RAF officer who later became a leading British forensic psychologist, drew up a list of characteristic features such as mode of dress and likely possessions as an aid to screening people suspected of being high-ranking Nazis.

WHOEVER FIGHTS MONSTERS

It was some years before Dr. Brussel's pioneering work was formally accepted by the law enforcement community. In 1970, after discussions with several psychiatrists, FBI agent Howard Teten began to give lectures at the FBI National Academy in Washington, D.C. The course, called "Applied Criminology" but nicknamed "Psych-crim," was based on unsolved cases from the previous seven years. In the course of expanding his presentations to local police departments, Teten was joined by Pat Mullany, from the FBI's New York office. Together they began to draw offender profiles in their classes.

The newly expanded FBI Academy opened at Quantico, Virginia, in 1972, and a Behavioral Science Unit (BSU) was set up there. After reading Dr. Brussel's *Casebook of a Crime Psychiatrist*, Teten visited him on several occasions to discuss the similarities and differences in their methods.

"His approach was to seek specific areas of psychiatric potential, and then combine them to form a profile. This was somewhat different from my approach, which was to derive an overall impression of the gross mental status, based on the crime scene as a whole. We reasoned that his method was more capable of providing detailed information, while my approach was less subject to error."

A CHILD DISAPPEARS

The fledgling BSU quickly had the opportunity to put their theories to practical test. In June 1973, seven-year-old Susan Jaeger was abducted from a tent in which she and her family from Michigan were camping, near Bozeman, Montana. Teten and Mullany compiled a preliminary profile of the likely perpetrator. He was, they decided, a young white male living in the area, who had come across the tent during a walk in the middle of the night. They concluded that Susan was probably dead – although her family continued to hope, against the odds, that she was still alive somewhere.

The buildings of the FBI Academy at Quantico, Virginia, opened in 1972. The fledgling Behavioral Science Unit was established in the windowless basement.

The local FBI agent, Peter Dunbar, had a suspect who fit this description: a 23-year-old Vietnam veteran named David Meierhofer, whom he described as "well groomed, courteous, exceptionally intelligent…." There was, however, no material evidence to connect him with the crime.

Then, in January 1974, a young woman who had rejected Meierhofer's advances disappeared, and he was again a suspect. Meierhofer volunteered to take a polygraph lie-detector test and be given "truth serum." He passed interrogation in both cases, and his attorney pressed for the authorities to cease harassing him.

Nevertheless, the Quantico team knew that such tests usually worked only for normal people. Psychopaths are able to separate the out-of-control personality that commits the crime from their more in-control selves. The in-control self suppresses all knowledge of the crimes committed, and so passes the tests. Teten and Mullany believed that the killer might well be the type who later telephones the relatives of his victims in order to re-experience the excitement of the crime. Dunbar therefore advised Mr. and Mrs. Jaeger to keep a tape recorder beside their telephone just in case. Sure enough, on the anniversary of the abduction, Mrs. Jaeger received a call from a man who said Susan was still alive. He said he had taken her to Europe, where he would give her a better life than the Jaegers could have ever afforded. "He was very smug and taunting," reported Mrs. Jaeger. "My reaction was not what he was expecting. I felt truly able to forgive him. I had a great deal of compassion and concern, and that really took him aback. He let his guard down, and finally just broke down and wept."

The call could not be traced, however. An FBI voice analyst concluded that the voice on the tape matched Meierhofer's, but this was not considered sufficient evidence in Montana to obtain a warrant to search the suspect's home.

Mullany felt that Meierhofer might be affected by a confrontation with Mrs. Jaeger, and arranged for her to meet him at his lawyer's office. During the meeting, he was calm and collected, in control of his emotions, but shortly after she had returned home to Michigan she received a collect call from a "Mr. Travis in Salt Lake City." He said it was he who had abducted Susan, but before he could continue, Mrs. Jaeger, recognizing the voice said, "Well, hello, David."

On this evidence, backed by an affidavit from Mrs. Jaeger, Dunbar was able to obtain a search warrant, and the remains of both missing girls were discovered in

All trainee FBI agents must undergo extensive weapons training. John Douglas, one of the first psychological profilers, was renowned throughout the Bureau as a crack shot.

The badge of an FBI agent. The Federal Bureau of Investigation grew from a division of the Justice Department that was originally formed in 1908. In 1924, J. Edgar Hoover was appointed director and remained in charge until his death in 1972. It was only subsequently that psychological profiling was adopted by the FBI.

Meierhofer's home. He readily confessed to the two murders and also to the unsolved killing of a local boy. Arrested, he hanged himself in his police cell the next day.

THE TEAM EXPANDS

By this time, the teaching and profiling case load for Teten and Mullany had become very heavy. Several other instructors, including Robert Ressler, Roy Hazelwood, and Dick Ault, joined them. Each in due course concentrated on a specialty, and Ressler eventually decided to conduct interviews with convicted serial killers. At that time, much of the teaching at Quantico was devoted to "canned" cases – those that had not been successfully solved – but Ressler had lectured principally on solved cases where the basic facts came from sources, such as newspaper articles and books, available to the public. Among these was the case of Harvey Glatman.

THE FATAL PHOTOGRAPHER

Harvey Murray Glatman equated sex with bondage. As a youngster in Boulder, Colorado, he enjoyed auto-erotic "games," hanging himself from a beam in the attic of his home to achieve orgasm. The family physician assured his parents that he would "grow out of it," if they could find a way "to keep him busy."

Harvey was highly intelligent – his IQ was later assessed at 130 – but unattractive to girls: he was an ugly teenager, with jug-handle ears and a weak chin. His idea of fun was to snatch a purse from a good-looking girl, then run away laughing before tossing it back to her. In 1945, Glatman threatened a girl with a toy gun and ordered her to strip; she screamed and he fled, but was quickly arrested. Released on bail, he headed east, and was again arrested, in New York City, after a series of robberies – this time with a real gun – and sentenced to five years in Sing Sing.

On his release in 1951, he returned to the west, where he set up a TV repair shop in Los Angeles. He took up photography and for some years led an outwardly quiet bachelor life – his lust fueled by glossy "glamour" magazines and the opportunity to observe models in a photographic studio.

On July 30, 1957, he made a TV service call at the apartment of a 19-year-old model. She had recently arrived in Los Angeles from Florida. In conversation, he told her he had been commissioned to take a series of photographs for a New York true-detective magazine – "the typical bound-and-gagged stuff." Two days later, he appeared at her door and took her to his home. There she permitted him to tie her hands behind her and gag her; after taking several photographs, he stripped her and, holding a gun to her head, raped her.

The couple sat together, naked, and watched television for a while. When the young woman promised Glatman that she would not report the rape if he let her go, he drove her 125 miles east of Los Angeles, into the desert near Indio. Once there, he again photographed her in her underwear, strangled her with a piece of rope, and buried her body in a shallow grave. It was so shallow that not long afterward the desert wind blew away the sand to reveal her remains.

Glatman then joined a "lonely hearts" club, using the name "George Williams," where he found his second victim. On March 8, 1958, he once again drove his victim into the desert, this time 55 miles away to Anza State Park, east of San Diego. There he raped her repeatedly at gunpoint, tied her up and photographed her, then strangled her with the same piece of rope. He left her corpse behind a cactus clump – but took away her red underwear as a souvenir.

"I DIDN'T WANT TO KILL HER."

The third victim was a part-time stripper who advertised in the personal columns of the *Los Angeles Times* that she was looking for modeling work. On July 23, 1958, Glatman went to her apartment with his gun and raped her; then he drove her out to the desert, where he spent all day alternately photographing and raping her. He almost let her go: "She was the only one I really liked," he later said. "I didn't want to kill her. I used the same rope, the same way." Realizing that the classifieds were a ready source of potential victims, Glatman began to place advertisements

One of the many photographs that Glatman took of his third victim, stripper Ruth Mercado. He carried her off to the desert near San Diego, California, where he tied her up and spent all day alternately photographing and raping her, before killing her.

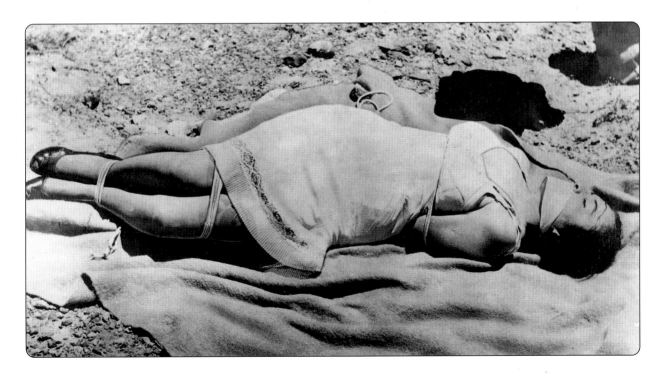

in newspapers, promising auditions for modeling opportunities for young women with no previous experience.

Los Angeles homicide detective Lieutenant Pierce Brooks was put in charge of the investigation of two of these apparently unconnected murders, but he felt they could well have been committed by the same man, who might also have committed more. On his own initiative, he spent weeks combing through newspapers and police files, looking for other killings with the same MO. He had begun to assemble telling evidence, when Glatman

attempted his fourth and final homicide.

After several failed efforts to lure hopeful models to his "studio," Glatman found one who was short of money and needed work. He picked her up in his car and set off for the desert along the Santa Ana Freeway; however, on the outskirts of Los Angeles, near the town of Tustin, she became alarmed. Glatman pulled over onto the hard shoulder, brought out his gun, and ordered her to strip.

"I knew he was going to kill me," she told police later. "I tried to plead, but I knew pleading wouldn't do any good."

Far left: Harvey Glatman's photograph of his first victim, 19-year-old Hollywood model Judy Dull, taken in his apartment. After stripping and raping her, he drove her to the desert near Indio, where he killed her and buried her in a shallow grave.

Jug-eared 30-year-old Harvey Glatman, following his arrest by sheriff's officers in Orange County. His fourth intended victim had seized his gun, and a passing patrolman had successfully rescued her.

A manacled Glatman stands, with San Diego Sheriff's officers and the deputy coroner, over the bones of his second victim, 24-year-old divorcee Shirley Bridgeford. He had left her body in Anza State Park, east of San Diego.

She grabbed at the gun: it went off and she was hit in the thigh, but she held on, leveling the gun at her abductor. He lunged for her, and together they fell from the vehicle, wrestling furiously for possession of the gun. It was at that moment that a police car drew up, and the patrolman was able to arrest Glatman.

At the police station, Lieutenant Brooks, the investigating officer, obtained an extensive confession from the killer. It was one of the first full documents of a serial killer's mentality, since Karl Berg's study of Peter Kürten. Glatman described how, after he had killed his first victim, he was overcome with remorse, and begged her corpse to forgive him. After some time, however, his feelings of fear and loathing were overcome by the memory of the excitement and release that he had enjoyed; each time he killed again, the compulsion grew stronger. But the guilt remained, and at his trial he begged for the death penalty. "It's better this way," he explained. "I knew this was the way it would be." He went to the gas chamber in San Quentin in August 1959.

FIGHTING MONSTERS

The Glatman case became an important case study for the FBI to test their criminal profiling methods. Ressler later wrote, in his book *Whoever Fights Monsters*:

"All we had to go on was what everyone else had. On Richard Speck, a killer of eight nurses in Chicago, the material was somewhat better, a book written by a psychiatrist who had done extensive interviews with him. Even these interviews were inadequate, though, because the man who had conducted them did not have the background in dealing with criminals, or the need to understand matters from a law-enforcement perspective, that would be necessary for our students. I wanted to better understand the mind of the violent criminal...."

In principle, the FBI had up to now shown little interest in murderers, rapists, child molesters, and similar cases of violent behavior, because these cases fell within the jurisdiction of local law-enforcement, and were not violations of Federal law. Despite the reservations of his colleagues, Ressler paid his dues as an associate of the American Psychiatric Association, the American Academy of Forensic Sciences, the American Academy of Psychiatry and Law, and others. "The Bureau's avoidance of mental-health professionals was of a piece with the Bureau's belief that if there was something worth knowing about criminals, the Bureau already knew it."

In the course of his wide reading (which included many confidential police files), Ressler came across a quotation from the German philosopher Friedrich Nietzsche's *Thus Spake Zarathustra* (1883–91), which he took as his guiding principle, and which he later used as the title of his first book:

"Whoever fights monsters should see to it that in the process he does not become a

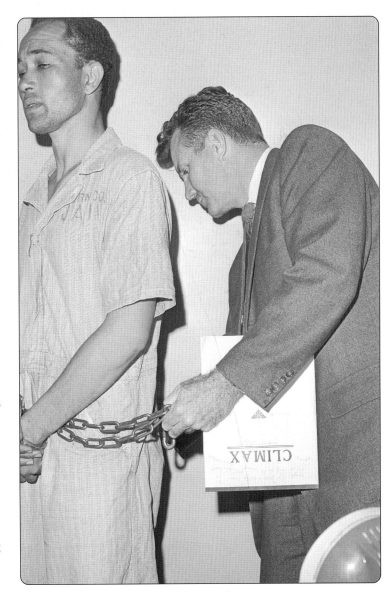

Pierce Brooks, who was a major influence in the establishment of the Violent Criminal Apprehension Program (VICAP), removes the manacles from Jimmy Lee Smith, a murder suspect.

At his trial, Harvey Glatman begged for the death penalty: "It's better this way," he said. He was executed in this gas chamber in San Quentin in August 1959.

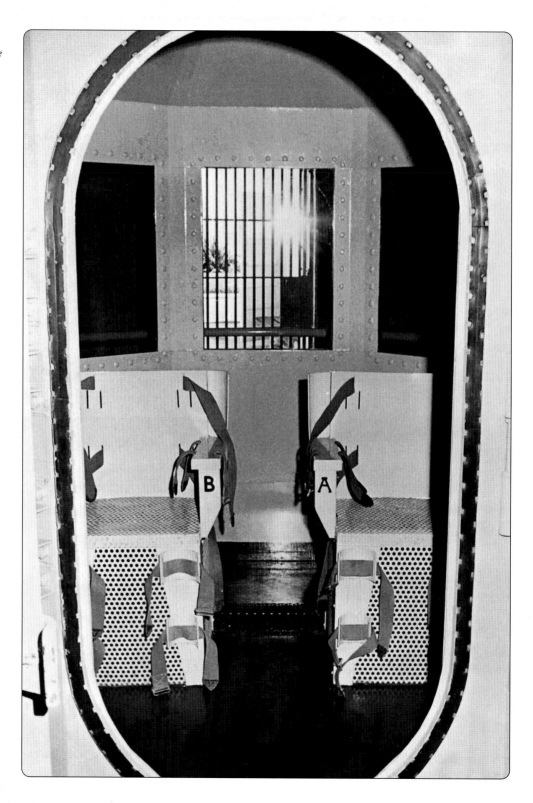

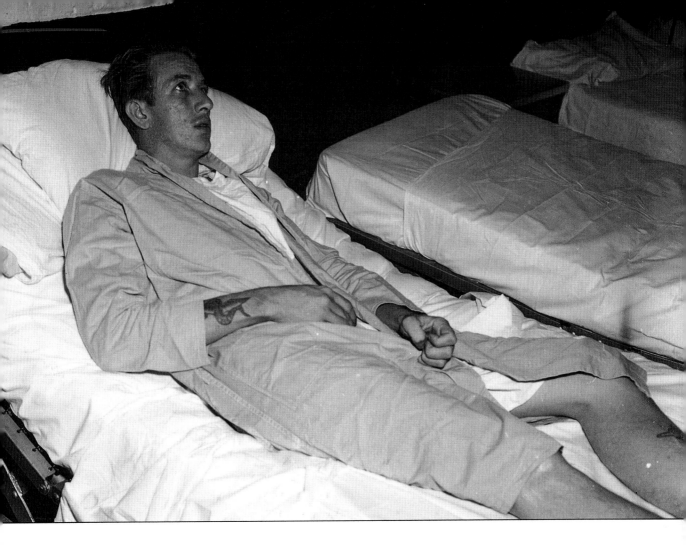

monster. And when you look into the abyss, the abyss also looks into you."

Ressler realized that dealing with monsters was tricky: he had to maintain a dispassionate attitude and understand – but not identify with – the thought processes of the criminal in order to develop an effective but impartial analysis.

CASE STUDY: RICHARD TRENTON CHASE

On January 23, 1978, a Sacramento truck-driver arrived at his home to discover his pregnant young wife butchered in their bedroom. An empty yogurt carton beside the body showed evidence that it had been used to drink the blood of the victim, and several pieces of her body were missing. There was no apparent motive for the killing.

The police contacted the local coordinator for the FBI's Behavioral Science Unit, Russ Vorpagel, and he in turn contacted Robert Ressler at Quantico. As it happened, Ressler was about to leave for the West Coast for a program of lectures, and before leaving he drew up a preliminary profile of the probable killer (see box, page 58).

Ressler made other conclusions that he did not include in this profile. If the killer possessed a car, it would be "a wreck, with fast-food wrappers in the back, rust throughout…." He also thought it likely

On the evening of July 14, 1966, Richard Speck (in jail, pictured above) murdered eight student nurses in their Chicago residence. Psychiatrist Dr. Marvin Ziporyn interviewed Speck twice a week over a period of six months, and he published his account, **Born to Raise Hell,** *in 1967. This book was among those studied by Robert Ressler.*

RESSLER'S PRELIMINARY PROFILE

"White male, aged 25–27 years; thin, under-nourished appearance. Residence will be extremely slovenly and unkempt, and evidence of the crime will be found at the residence. History of mental illness, and will have been involved in use of drugs. Will be a loner who does not associate with either males or females, and will probably spend a great deal of time in his own home, where he lives alone. Unemployed. Possibly receives some form of disability money. If residing with anyone, it would be with his parents; however, this is unlikely. No prior military record; high school or college dropout. Probably suffering from one or more forms of paranoid psychosis."

An artist's drawing of Richard Trenton Chase. It illustrates the accuracy of Robert Ressler's preliminary profile.

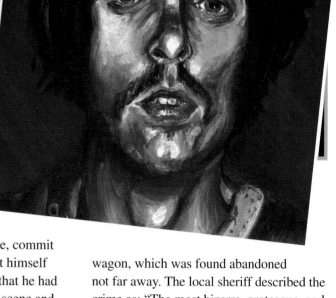

that the killer lived near the victim "because he would probably be too disordered to drive somewhere, commit such a stunning crime, and get himself back home." Ressler thought that he had walked to and from the crime scene and guessed that he had been let out of a psychiatric-care facility not much more than a year earlier.

But even as Ressler was packing his bags for California, the murderer struck again. On January 26, three bodies were discovered by a neighbor, in a home within a mile of the previous slaughter. They were a 36-year-old woman, her six-year-old son, and a male family friend. All had been shot with a .22, the woman mutilated, and her 22-month-old nephew abducted. The killer had apparently left in the friend's station

wagon, which was found abandoned not far away. The local sheriff described the crime as: "The most bizarre, grotesque, and senseless killings I've seen in 28 years."

Ressler added more details to the profile he had prepared: "single, living alone in a location within one-half to one mile from the abandoned station wagon." He told Vorpagel that "before this man had murdered, he had probably committed fetish burglaries in the area and, once he was caught, we'd be able to trace his crimes and difficulties back to his childhood." He characterized "fetish" burglaries as those in which the items stolen or misused are articles of women's clothing, rather than jewelry or other items of marketable value, which the burglar takes for auto-erotic purposes.

Acting on this information, more than 65 police officers went out to question hundreds of people within a half-mile radius of where the station wagon had been found. A vital lead came from a woman in her late twenties, who had spoken to a man she had known in high school 10 years earlier, named Richard Trenton Chase. She said she had been shocked at his appearance: "disheveled, cadaverously thin, bloody sweatshirt, yellowed crust around his mouth, sunken eyes."

CHASE IS ARRESTED

There were other suspects, but two police officers staked out Chase's home, which was less than a block away from the site of the abandoned station wagon. When he emerged, carrying a box under his arm, they arrested him in possession of a .22 revolver and the wallet of one of his victims. His truck was parked nearby. A dozen years old, it was in poor condition and littered with discarded newspapers, rags, beer cans, and empty milk cartons. There were also a 12-inch butcher's knife and rubber boots spattered with blood. The box he was carrying was packed with bloodstained rags.

Inside Chase's apartment – a slovenly mess – were three food blenders with blood in them, dirty and bloody clothing, and a number of newspaper articles about the first murder. The refrigerator held several dishes with body parts in them, as

well as a container with brain tissue. A calendar on the wall was marked "Today" on the dates of the two sets of killings – and, ominously, the same inscription appeared on 44 more dates throughout 1978.

After his arrest, Chase's horrific history emerged. Born in 1950, he had exhibited psychotic behavior from an early age; later he became a heavy drug user, and had been arrested for possession of marijuana. When he went to live alone in an apartment, he

Robert Ressler was one of the first FBI criminologists to develop criminal profiling techniques.

Three very different images of Charles Manson, who inspired the hippie followers of his "Manson family" to carry out a succession of motiveless killings in Los Angeles in August 1969. Among their victims was Sharon Tate, the actress wife of movie director Roman Polanski.

Right: The arrest of Sirhan Sirhan, who was found guilty of the assassination of Senator Robert Kennedy in Los Angeles on June 5, 1968.

began to kill and disembowel rabbits and drink their blood "to stop my heart from shrinking." Eventually he was committed as a schizophrenic to a psychiatric institution. There he earned the nickname of "Dracula" when he killed two birds, and was found with blood around his mouth. After a course of medication, he was released to the care of his parents, and his mother paid the rent on his apartment and bought his groceries.

However, Chase acquired guns, and began to kill neighbors' pets and drink their blood; his progression to human killing was inevitable, and it was fortunate that he was caught early in his planned orgy of murder.

Robert Ressler's profile had been strikingly accurate, but, as he later wrote: "Profiles don't catch killers, cops on the

beat do.… My profile was an investigative tool, one that in this instance markedly narrowed the search for a dangerous killer. Did my work help catch Chase? You bet, and I'm proud of it. Did I catch him myself? No."

"At last," continued Ressler, "I came to the point where I wanted very much to talk to the people about whom I'd been lecturing, the killers themselves." A new recruit to the Quantico instructors was John Douglas, who accompanied his senior colleague to California. Ressler enlisted the assistance of the FBI liaison officer to the state prison system; together, within a few days, the three men interviewed seven of the most notorious convicted killers, including Sirhan Sirhan (the assassin of Robert Kennedy), Charles Manson, and Edmund Kemper.

KEMPER'S THOUGHTS

"I thought of making this a demonstration to the authorities in Santa Cruz – how serious this was, and how bad a foe they had come up against.... I had thought of annihilating the entire block that I lived on.... Not only the block that I lived on but the houses approaching it, which would have included as many as ten or twelve families. And it would be a very slow, a very slow, quiet attack....

"I had done some things, and I felt that I had to carry the full weight of everything that happened. I certainly wanted for my mother a nice, quiet, easy death like I guess every one wants."

CASE STUDY: EDMUND KEMPER

Edmund Emil Kemper was born in December 1948, in Burbank, California. He grew up with two younger sisters in a family where his mother and father constantly fought, and eventually separated. Death and execution fascinated him from an early age. At the age of 14, he was described by his mother as "a real weirdo." He was sent to stay with his grandparents on their isolated California ranch, where he shot both of them in August 1964: "I just wondered how it would feel to kill grandma," he said.

Kemper, diagnosed with "personality trait disturbance, passive-aggressive type," was committed to the care of Atascadero State Hospital for the criminally insane; however, in 1969, at age 21, he was released into the custody of his mother – despite the objections of the state psychiatrists. He was a giant figure, six feet nine inches tall and weighing nearly 300 pounds. He bought a car and roamed the highways, at first only giving lifts to young female hitchhikers. Then, in May 1972, he held two co-eds from Fresno State College at gunpoint, took them into an isolated canyon and stabbed them to death, before carrying their bodies back home. There he decapitated and dissected them, taking Polaroid photos of his actions, before burying the remains in the Santa Cruz Mountains – but he kept their heads for some time before disposing of them.

Four months later, a 15-year-old Japanese student suffered a similar fate. Another four months passed before Kemper's next kill, a student at Cabrillo College. When police discovered her dismembered body some days afterward, Kemper still had her head in a box in his closet; later he buried it in his mother's backyard. With a macabre sense of humor, he set the head facing his mother's bedroom window: She had always said that she wanted people "to look up to her."

Kemper's activities now began to accelerate. In February 1973 he picked up two female hitchhikers on the campus of Santa Cruz's Merrill College (where his mother was an administrator), and almost immediately shot both. When he arrived at his mother's home she was there, and he was obliged to carry out his decapitations in the trunk of his car. The next morning he washed the blood from the bodies before dumping them in an isolated canyon nearby.

At dawn on Easter Saturday Kemper killed his mother with a hammer, crushing her skull, and then cut off her head. As a final touch, he cut out her larynx and fed it into the garbage disposal: "It seemed appropriate," he later confessed, "as much as she'd bitched and screamed and yelled at me over so many years."

He then invited his mother's close friend to a "surprise" meal. When she arrived and said, "Let's sit down, I'm dead," he took her at her word and strangled her, decapitated her, and drove off in her car. Early the next morning, not knowing what he should do, he drove aimlessly eastward for many hours, having left a note for the police in his mother's home:

"No need for her to suffer any more at the hands of this horrible 'murderous Butcher.' It was quick – asleep – the way I wanted it. Not sloppy and incomplete, gents. Just a 'lack of time.' I got things to do!!!"

At last, in Pueblo, Colorado, he pulled over to a roadside telephone booth, called the Santa Cruz police, and told them he was the "Co-ed Killer." They did not believe him,

and it took several similar calls before the Colorado police arrived to arrest him.

At his trial, Kemper explained his motive in killing young women:

"Alive, they were distant, not sharing with me.... When they were being killed, there wasn't anything going on in my mind except that they were going to be mine.... That was the only way they could be mine."

Edmund Kemper on his way to court in Pueblo, Colorado, after he confessed to police that he had killed his mother, his mother's friend, and six other female victims.

He was judged sane and found guilty on eight counts of murder. He asked for "Death by torture," but was sentenced to life imprisonment, without the possibility of parole.

THE "CO-ED KILLER"

Douglas later wrote his impressions of this "Co-ed Killer," whose IQ had been assessed at 145. He found him "neither cocky and arrogant, nor remorseful and contrite." Kemper was cool and soft-spoken, and the only times he showed emotion were when he recalled how badly his mother had treated him. Nevertheless, when he had expressed an interest in joining the California Highway Patrol, she had done her best to have his juvenile record for murder expunged.

This interest in police work, Douglas subsequently realized, was a common factor in many serial killer cases. Frequently, such murderers drove a police-like vehicle, or even a decommissioned police car. Kemper related how he had frequented bars and restaurants known to be popular police resorts, and would strike up friendly conversations. This not only made him feel like an insider, a near-policeman himself, but it also kept him informed of the progress of investigations into his killings.

Kemper had learned a great deal about psychology during the five years he had already spent in prison, and he could produce a precise psychiatric analysis of his behavior. He provided details of how he picked up his victims without arousing suspicion. When he stopped his car for a pretty girl, he asked her where she was going, then looked at his watch – as if deciding whether he had sufficient time. This would immediately put the girl at ease, because a busy man would not be cruising for pick-ups. Douglas reasoned that "the normal common-sense assumptions, verbal cues, body language, and so on, that we use to size up other people, and make instant judgments about them, often don't apply to sociopaths."

After several long interviews over the years with Kemper, Douglas wrote: "I would be less than honest if I didn't admit that I liked Ed. He was friendly, open, sensitive, and had a good sense of humor: but my personal feelings about him…do point up an important consideration for anyone dealing with repeat violent offenders. Many of these guys are quite charming, highly articulate, and glib."

The conclusion reached was that Kemper's killings were the expression of a fantasy to be rid of his domineering mother. She had told him that he could never hope to be married to any of the attractive co-eds that he fancied, and he took her at her word.

DEVELOPING THE TECHNIQUE

On their return to Quantico, Ressler and Douglas took every opportunity to interview other killers in detention. After an internal disciplinary inquiry into these unauthorized interviews, they were exonerated, and soon afterward the FBI hierarchy approved Ressler's Criminal Personality Research Project. Ressler now had to convince the police that profiles of unknown criminals were a valuable tool of investigation. The FBI team soon had success using their profiles with a case that stymied the police in New York City.

CASE STUDY: CARMINE CALABRO

One afternoon in October 1979, the naked body of young Francine Elverson was discovered on the roof of the building in the Bronx where she lived with her mother and father. She had not been seen since early that morning, when she had left for her job as a teacher of handicapped children at a nearby day-care center. Homicide detectives decided that she had been battered unconscious as she walked downstairs, and then carried to the building's roof.

Her killer had neatly placed her earrings on either side of Francine's head; her nylon stockings were tied loosely around her wrists, and her underwear had been taken off and placed over her head to hide her face. The rest of her clothing was piled close by, and beneath it was clear evidence that her killer had defecated. The young woman had been beaten around her face, strangled with the strap of her purse, and brutally mutilated when she was dead. Her nipples had been cut off and then placed on her chest, there were bite marks on her thighs and knees, she had been sexually assaulted with her umbrella and a pen, and a comb was twisted into her pubic hair. Across her abdomen and onto her thigh the killer had scrawled in ink: "Fuck you. You can't stop me."

> "I would be less than honest if I didn't admit that I liked Ed. He was friendly, open, sensitive, and had a good sense of humor."

Far left: The immense size of Edmund Kemper – six feet nine inches tall, and weighing nearly 300 pounds – is strikingly revealed in this photograph, as he is escorted into court by law-enforcement officers.

A typical view of the Bronx during the 1970s. The body of Francine Elverson, strangled with the strap of her purse, was found on the roof of a Bronx apartment building in October 1979.

Perhaps strangest of all, Francine's body had been unusually arranged. Her parents told investigators that the arrangement resembled the Hebrew letter *chai*, an emblem that had been on a chain she wore around her neck, and that was missing from the crime scene.

At the autopsy, traces of semen were found on Francine's body, together with a single black pubic hair that was not hers.

A team of 26 New York police questioned more than 2,000 people and assembled a list of 22 suspects. The list included a man living in the building who had previously been incarcerated for sexual offenses, a man who had formerly been a janitor, and a 15-year-old boy. The boy said he had found Francine's wallet on the stairs on his way to school and had not handed it to his father until his return. After a month, the

THE FBI PROFILE

The FBI team decided that this was a crime of opportunity, a spontaneous event. They described a white male, aged 25 to 35, who knew the victim and lived in the building, or close by, with his parents or an older female relative. He would be disheveled in appearance – although not a drug or alcohol abuser – unemployed or working at night, and did not own a car or hold a driver's license. He was mentally ill (because this was not a premeditated killing) and "the disease had been bubbling within him for 10 years before it erupted into a mutilation murder." This was his first serious crime, but was unlikely to be his last.

The killer had probably been released from a mental-care institution within the past year, and was still taking medication. The scribbled message and the arrangement of the body suggested that the killer was probably a school dropout who had obtained his ideas from a large collection of pornographic literature.

"You don't have to go far for this killer," Douglas told the investigators. "And you've already talked to this guy." Robert Ressler, who was also consulted, concluded, despite disagreement with some of his colleagues, that the black pubic hair was irrelevant to the investigation.

investigation was stalled. The two detectives in charge of the case took all the relevant files, including crime scene photos, to the FBI. There, a team of four, including John Douglas and Roy Hazelwood, examined the evidence.

The profile provided by the FBI persuaded the police to turn their attention away from many of their suspects and focus upon another possibility: 30-year-old Carmine Calabro, an unemployed stagehand who lived with his father and whose mother had died 11 years previously. The father said his son had been in a secure mental institution at the time of the murder.

The police decided to look more closely into Carmine's alibi. He had spent more than a year in a nearby institution, but it soon emerged that he could have slipped out and subsequently returned without anyone having noticed. His room at the apartment was full of pornographic material. At the time of the killing he had been wearing a plaster cast on his arm –

which had since been discarded – and it was surmised that he had used this to beat his victim unconscious. The bite marks on the victim were matched conclusively by three forensic odontologists to Carmine's teeth, and the case was solved.

As for the single black hair, it turned out that Francine's body had been carried to the medical examiner in a body bag that had not been properly cleaned. Police Lieutenant Joseph D'Amico later related: "They had him [Calabro] so right that I asked the FBI why they hadn't given us his phone number, too."

By this time the BSU had accumulated data on a large number of cases of repetitive violent behavior. Police departments began to send in details of unusual cases for analysis, and it became necessary to develop a classification system – relatively free of psychiatric jargon – that could be exploited to explain the different types of offenders to them. Until now, law enforcement had to rely largely on the standard publication

Diagnostic and Statistical Manual of Mental Disorders (DSM) for guidance, and the FBI researchers found it of little value in their work. This was the starting point for the preparation of the FBI *Crime Classification Manual,* which was eventually published in 1992. John Douglas wrote: "With *CCM* we set about to organize and classify serious crimes by their behavioral characteristics and explain them in a way that a strictly psychological approach such as *DSM* has never been able to do. For example, you won't find the type of murder scenario of which O. J. Simpson was accused in *DSM*. You will find it in *CCM*."

The Army also asked the FBI to provide training sessions, principally in hostage negotiation, so Robert Ressler and John Douglas traveled to Germany to teach at military bases. On their way back to the United States they were invited to visit Bramshill, the British Police College. This was at a time when a huge force of detectives was engaged in the hunt for the "Yorkshire Ripper," and the topic naturally arose during after-hours discussions.

The Behavioral Science Unit's stated mission "…is to develop and provide programs of training, research, and consultation in the behavioral and social sciences for the FBI and law enforcement community that will improve or enhance their administration, operational effectiveness, and understanding of crime."

THE YORKSHIRE RIPPER

He was described by his wife as a caring man, careful and conservative in his habits, but beneath his innocent-seeming façade raged the deep implacable hatred of one of the most notorious serial murderers of the 20th century.

His first victim, murdered on October 30, 1975, was a part-time prostitute in the city of Leeds, Yorkshire. He killed her with several blows of a ball-headed hammer to her skull and then, in what was to become his "signature," stabbed her 14 times in the torso and abdomen with a sharpened cross-head screwdriver. At first police considered this to be a typical example of the hazards faced by prostitutes, but when a second prostitute was found with similar injuries – and 50 frenzied stab wounds all over her body – they recognized that the assailant was likely to be the same person. However, they failed to realize, for more than three years, that the attacks were related to two earlier attacks upon women – neither of them prostitutes.

On May 9, 1976, a man with a dark beard attacked a young woman. She screamed, and he ran off. Nearly a year passed before the body of another victim was discovered: she had been struck down with three blows of a ball-headed hammer and stabbed more than 20 times with a screwdriver and a snap-blade knife. It was then, in recollection of the crimes of "Jack the Ripper," that the killer earned the name of the "Yorkshire Ripper."

There was near panic in the red light district of Chapeltown in Leeds. Some women moved far away, to London, Manchester, or Birmingham, but many set up only a few miles distant, in the nearby city of Bradford. The Ripper followed

Police experts examining one of the victims of the Yorkshire Ripper, who terrorized the red-light districts of Leeds and Bradford in England from 1975 through 1980.

them: on April 23, 1977, the body of another prostitute, similarly killed and mutilated, was found in her blood-soaked bed. Two months later, a 16-year-old girl, a respectable store employee in Leeds, was murdered in a similar manner.

The killings continued: in Manchester, nearly 40 miles from Leeds, as well as in nearby Huddersfield and Bradford again. By now more than 200 police officers had been assigned to the investigation, but they came no closer to finding the Ripper. Then, for nearly a year, the murders ceased. In 1979 they began again, and shortly after the first renewed attack, on April 4th, the Chief Inspector in charge of the investigation, George Oldfield, received two letters signed "Jack the Ripper," and later a two-minute tape-recording.

These communications diverted the course of the police inquiry. Voice experts identified the accent as coming from the Wearside region of northeastern England, so the tape was played at pubs and clubs in the area in the hope that someone might identify the voice. A toll-free telephone number was announced so that others might hear the recording, but after thousands of calls from the public the police were no closer to identifying a suspect. Exhausted, George Oldfield suffered a heart attack. He was replaced, and after many weeks of wasted effort, it was decided that, as in the case of Jack the Ripper (which they closely resembled), the

The discovery of another of the Yorkshire Ripper's victims, struck down with a ball-headed hammer and stabbed many times with a sharpened screwdriver.

messages were a hoax, perpetrated by a disgruntled police officer who had a grudge against Oldfield.

Meanwhile, the killer no longer pursued only prostitutes; he targeted any lone woman he might meet in the night – a 19-year-old clerk, a university student, and a middle-aged civil servant. In October and November 1980, two women survived his attacks. He killed again in mid-November, his victim another young university student. But she was his last.

On January 2, 1981, the killer picked up a prostitute in the city of Sheffield. A few minutes later, his car was stopped by a police patrol car, and a quick computer check established that the registration plates were stolen. After his subsequent arrest, the dark-bearded man gave his true

THE RIPPER TAPE

"I'm Jack. I see you are still having no luck catching me. I have the greatest respect for you, George, but you are no nearer catching me now than four years ago when I started. I reckon your boys are letting you down, George; ya can't be much good, can ya? The only time they came near touching me was a few months back, in Chapeltown when I was disturbed. Even then it was a uniformed copper, not a detective. I warned you in March that I'd strike again…but I couldn't get there. I'm not quite sure when I'll strike again but it will definitely be [some] time this year, maybe September, October, even sooner if I get the chance: there's plenty of them knocking about. They never learn, do they, George…. I'll keep on going for quite a while yet. I can't see myself being nicked just yet. Even if you do get near, I'll probably top myself first. Well, it's been nice chatting to you, George…."

Truck driver Peter Sutcliffe, finally arrested after a murderous career that lasted nearly six years.

name of Peter Sutcliffe; the following day, a hammer and screwdriver that he had dumped were recovered, and a search of the police station where he was taken revealed a knife hidden in a cistern. When Sutcliffe was told of this, he said: "I think you are leading up to the Yorkshire Ripper.… Well, that's me."

At his trial, Sutcliffe pleaded manslaughter, but he was found guilty on 13 counts of murder and seven of attempted murder, and jailed for life.

Before Sutcliffe was arrested, the FBI men had asked if they could see crime scene photographs (which were not available), and the British officers expressed considerable skepticism that anything about the case could be discovered from them. However, a copy of the "Ripper tape" (believed at that time to have been recorded by the serial killer) was played, and at the end Ressler said: "You realize, of course, that the man on the tape is not the killer?" Douglas concurred: "Based on the crime scenes you've described, and this audiotape…that's not the Ripper. You're wasting your time with that."

SPEEDY
CONFIDENT
AND
OBSESSIVE

SUNDERLAND ACCENT
5' 10" TALL
WHITE AGE 30-45
STRONGLY BUILT
BLOOD GROUP AB
SINGLE AND
LIVING ALONE?

WORKING IN
YORKSHIRE?
USES WORK BENCH
EQUIPMENT TO
MUTILATE VICTIMS
WEARS HEAVY
INDUSTRIAL BOOTS

SKETCH BY
TIM HOLDER

A HISTORICAL PERSPECTIVE

The first serious study of the criminal classes was a prize-winning paper presented to the French Academy of Moral and Political Sciences in 1838 entitled "The dangerous classes of the population in big cities, and the means of making them better." This was followed 30 years later by Lombroso's recognition of different criminal types. English sociologist Dr. Charles Goring's counterclaim, in 1913, that "the one vital mental constitutional factor in the etiology of crime is defective intelligence" was supported by studies throughout the 1920s, and it was not until the end of the decade that attention shifted to the consideration of the disordered personality.

In 1932, the Psychiatric Clinic of the Court of General Sessions in New York state began to classify offenders by personality evaluation: presence or absence of psychosis; intellectual level; presence of psychopathic or neurotic features; and physical condition. A long-term project at Bellevue Psychiatric Hospital established that aspects of personality were the most significant in assessing criminality. Criminal behavior is not far removed from normal behavior, and psychologists concluded that it derived from three behavioral areas: the aggressive tendency, both destructive and acquisitive; passive, or subverted, aggression; and psychological needs.

Classification of criminals gradually became an important aspect in correctional facilities in the United States, and in 1973 the National Advisory Commission called for criminal classification programs to be initiated throughout the criminal justice system.

Challenged to provide an off-the-cuff profile, the two men said that the Ripper was not someone who would communicate with the police. He was an "almost invisible" loner, in his late 20s or early 30s, probably a school dropout or a man who had not been through higher education. He was able to enter the areas of the murders without attracting attention because his work regularly took him there: as a cab or truck driver, a postal worker – or even a policeman. He probably had a relationship with a woman, but he had serious mental problems that had taken years to develop, and his murders were an attempt to punish women in general.

When the Yorkshire Ripper, 35-year-old Peter Sutcliffe, was apprehended, he turned out to be a married truck driver who worked for an engineering firm. He had left school at 15 after a long record of truancy, and met his wife, 16-year-old Sonia Szurma, at the age of 21. The couple broke up several times in the seven years before they married; they often had furious arguments, and it was within a year of his marriage that he committed his first murder.

In a mammoth 17-hour recorded confession he stated that his hatred of prostitutes had begun years earlier in 1969, when he was 22 years old, and a streetwalker, whom he had picked up, had refused to give him change for a £10 note he had handed her.

Far left: A collection of Identikit pictures and artists' impressions of the Yorkshire Ripper. Peter Sutcliffe was questioned three times by police and each time released, despite his resemblance to several of these pictures.

British police marked off the backyard of a North London house in 1983 in the search for the buried bodies of as many as 13 young men murdered by Dennis Nilsen. Nilsen – a good example of an organized offender – calmly described his killings to the police after his arrest.

ORGANIZED AND DISORGANIZED PERSONALITIES

The first step in the FBI classification of violent criminals was to divide them into "organized" and "disorganized" personalities. This is a generalization and, as with all generalizations, there are exceptions: the BSU soon found that they had to introduce a third personality type, which they called "mixed."

The leading characteristic of the organized killer is his planning of the crime; it is premeditated, not committed on the spur of the moment. The planning forms part of the offender's fantasies, which have probably been dwelled on for years before they find violent expression. The victims are mostly strangers, of a particular type that the offender has in mind, and that he has been hunting for. Since the crime has been planned, the offender will have figured out ways to approach the victim, win their confidence, and so gain control over them.

The disorganized killer doesn't choose victims logically. He may often, in error, pick upon a victim who is not easily controlled, who may fight back, and whose body will be marked with defensive wounds on the hands and arms. He does not know, nor have any interest in, their identity and characteristics; this is often evident in the covering of their faces, or extensive mutilation of their features.

Apart from planning, organized killers can also show an intelligent ability to adapt to a changing situation. When Edmund Kemper abducted and shot two girls on the campus of Merrill College, he wrapped their bodies in blankets, placing one in the passenger seat and one in the back. He explained to security guards at the campus exit that both were drunk, and he was taking them home.

In addition, organized offenders learn as they progress; from crime to crime, they "improve" on what they do. If police report a series of murders with apparently the same MO, FBI profilers advise them to pay particular attention to the first, because it will probably have taken place closest to where the killer lives or works. With growing experience, he will likely leave the bodies of the victims farther and farther away from where he abducted them.

Another indicator of planning is the "rape kit" – handcuffs, rope, etc. – that the organized killer carries, generally in the

ORGANIZED AND DISORGANIZED OFFENDERS

The FBI established the following characteristics as being typical of organized and disorganized offenders.

ORGANIZED	DISORGANIZED
Average to above average intelligence	Below average intelligence
Socially competent	Socially inadequate
Likely to be a skilled worker	Unskilled worker
Sexually competent	Sexually incompetent
High in order of birth	Low in order of birth
Father's employment stable	Father's employment unstable
Inconsistent childhood discipline	Harsh discipline as a child
Controlled mood during crime	Anxiety during crime
Alcohol use associated with crime	Minimal use of alcohol
Precipitating situational stress	Minimal situational stress
Living with partner	Living alone
Mobility, with car in good condition	Lives/works near crime scene
Interest in news media reports of crime	Minimal interest in news media

Following the crime/s:

May change employment or leave area	Significant behavior change (for example, drug or alcohol use)

trunk of his car. He brings his own weapon to the scene and takes it away afterward. He may wipe away fingerprints, even clean up blood, making a conscious effort to avoid identification. Carrying away the body and concealing it is another attempt to delay identification, this time of the victim, who often is found naked and without any personal belongings. Decapitation is the most developed aspect of this intention.

The organized offender often keeps some of the personal belongings of his victim – wallets, rings and other jewelry, or articles of clothing – as "trophies," to be gloated over as he renews his fantasies. The disorganized killer is more likely to cut off a random part of the body or a lock of hair.

Organized serial rapists and murderers often carry a "rape kit," which will often include weapons, materials to bind their victims, and torture implements.

STAGING

A vital element in the analysis of violent crimes is what is known as "staging" – the way in which some killers attempt to make changes to the crime scene to mislead the investigators, or later take steps to divert the course of the investigation. Spouses, for example, having killed their mate in a violent fit of anger, may then ransack the home to make it look as if a burglar were responsible for the murder. Some killers can go further, often in extremely ingenious ways.

In the evening of a day in late February 1980, teenager Debra Sue Vine disappeared while walking the two blocks from a friend's home to her own in the small town of Genoa, Ohio. The next morning a telephone call – sounding as if it came from a white male in his early 20s, with a New England or Southern accent – announced: "We have your daughter. We want $80,000, or you will never see her again." The next day Debra's father received a call (which was recorded) from a man speaking with a Mexican accent who demanded $50,000 and said that further instructions would follow. Because these ransom demands had been made, the FBI was able to take up the case.

Three days later, some of Debra's clothing was found beside a country road, two miles from Genoa, and the rest was found on another road the following day. With it was a crumpled sheet of yellow legal paper on which a rough map had been drawn, with markings indicating a bridge over a nearby river. Police immediately went to the spot and found tire tracks and sets of footprints suggesting that someone had dragged something to the bridge and dropped it in. Convinced

THE CRIME SCENE

Details of the crime scene, particularly photographs, are of great value to a profiler. In 1992, the FBI produced their *Crime Classification Manual*, written by John Douglas, Robert Ressler, Ann Burgess of the University of Pennsylvania, and Allen Burgess of Northeastern University, with three additional contributors. Under the heading "Crime scene indicators," they list the following questions to be considered:

Has the crime occurred indoors or outside? At what time did it take place? Where did it take place? How long did the offender stay at the scene?

How many offenders were there?

Is the crime spontaneous and disarrayed, with physical evidence at the scene? Or does it indicate a methodical, well-organized perpetrator? The text emphasizes that the crime scene will seldom be completely organized or disorganized. It is more likely to be "somewhere between the two extremes of the orderly, neat crime scene and the disarrayed, sloppy one."

Did the offender bring the weapon to the crime scene? Or was it a weapon of opportunity, picked up at the scene? Is the weapon absent, or has it been left behind? Is there evidence of several weapons and traces of ammunition? Was the body left openly displayed, or placed in a way to make sure it was discovered? Or was the body concealed or buried? Did the offender seem to have little concern whether the body would be discovered or not?

The presence of items added to the scene, or the absence of others, is also important in classifying the offense.

Finally, the text asks how the offender controlled the victim. Are elements of something like a "rape kit" evident at the scene, or was the offender unprepared, attacking suddenly, and physically overpowering the victim?

Police search for clues following the disappearance of Chandra Levy in 2002. Physical evidence at the scene of the crime provides invaluable clues for the profiler.

that this was where Debra's body had been dumped, the police searched the river, but could find nothing.

Robert Ressler happened to be teaching in the vicinity, and he was given full details of the girl's abduction. Almost immediately, he concluded that the clues had been deliberately staged, and that it was necessary to look in the opposite direction from the one in which the investigators had been led. He then drew up a profile of the probable offender.

He would be a man in his late 20s to early 30s, and athletically built, because he was strong enough to abduct Debra on the street without difficulty – an antisocial

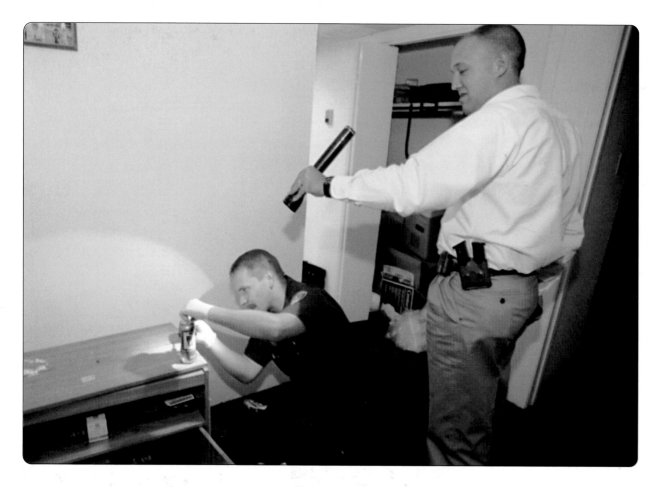

FBI agents search a burgled and wrecked apartment for clues. In rare cases, such damage may have been caused by the occupant, in the hope of making a false insurance claim.

personality who compensated by muscle-building and driving around in powerful cars. A macho individual, he would nevertheless be neat in appearance, and might have some reputation as a ladies' man. The staging showed that he was familiar with police methods, and Ressler suggested that he had previously been a police officer, a private detective, or a security officer, and drove a car that looked like a police vehicle, with a CB radio. Now unemployed for several months, he might in that time have been in trouble with the law, and arrested.

Two likely suspects fit this profile: a 31-year-old recently fired from the Genoa

police department; and a man who had served in a neighboring police department, then worked as a railroad detective in Michigan. The first was able to establish a firm alibi; however, the second could not, and he fit the profile particularly well. After being fired from the railroad, he had been arrested in Michigan for burglary. He was known to delight in telling Mexican jokes, and he drove a powerful car with a CB radio. Investigators decided to keep him under discreet surveillance.

It was several weeks before Mr. Vine received another call from the man with the Mexican accent. The call was traced to a Genoa payphone, and the next afternoon

FBI agents saw the suspect make a call from the same phone – at the same time Debra's father was being told by phone that he would receive instructions that evening. Then the suspect, wearing gloves, taped a folded note beneath the phone-booth shelf. Later, following a series of phone calls, Mr. Vine and the agents were led on a wild-goose chase from phone booth to phone booth, nine in all, in each of which a note had been taped. The evidence was sufficient for the man to be arrested, charged with extortion, and convicted. As Ressler had predicted, Debra's body was eventually found in the opposite direction from that indicated on the map, and the police pursued their inquiries in the hope of pinning her murder on the convicted offender.

> **"The woman's anger was directed at a close family member; she was seeking attention; and she hoped to obtain compensation."**

STAGING AN INSURANCE CLAIM

The detection of staging can also be of great importance in other types of crime. In 1991, shortly after Ressler had retired from the FBI and become a crime consultant, he was able to advise an insurance company on a claim for $270,000 worth of damage to a home, which was said to be the work of teenage vandals. Working solely from photographs of the scene, Ressler described "damage spread through the living room, halls, kitchen, main bedroom, and bath. Walls, furniture, paintings, clothing, vases, jade carvings, and other items had been broken and defaced. Curtains were down. Glass over art prints had been cracked. Spray-paint graffiti could be made out in various locations, on walls, furniture, and the like, in single-word clusters, such as 'Asshole,' 'Ass,' 'Suck,' and 'Cunt.' There was also a two-word inscription: 'Fuck me.'"

But the destruction was selective. Some paintings that did not look very valuable had been badly damaged, but their ornate frames were intact. The glass over the valuable prints was cracked, but the prints unharmed, and one large oil painting of a little girl was untouched. Vases and statues littered the floor, but were unbroken; doorknobs were damaged, but not the doors themselves, and the partition walls had not been kicked in. Most telling of all: a curtain rod had been taken down and laid gently on the floor, without harming the curtains.

As for the graffiti, Ressler decided they were untypical of teenage vocabulary; and "Fuck you" would have been more likely (as in the case of Carmine Calabro) if written by a hostile, arrogant young man. He concluded that the damage had been perpetrated by a lone white female, between 40 and 50 years of age. Probably the mother of an only daughter, she had gone through several divorces and suffered severe stress – through money or men problems, or the loss of her job – in the days before the event. The obscene graffiti reflected her ideas of how she fantasized male hostility. Summing up, he wrote that the woman's anger was directed at a close family member; she was seeking attention; and she probably hoped to obtain compensation to carry out renovations that she could not otherwise afford.

A psychologist who had been retained

Secret Service men wrestle John Hinckley to the ground after his unsuccessful attempt to assassinate President Ronald Reagan in Washington, D.C., on March 30, 1981.

by the insurance company reported that this profile exactly matched the woman who had made the claim. "White and in her 40s, she had broken up with her boyfriend, had money problems, had a daughter who lived with her former husband.… The psychologist was rather amazed at my perspicacity. I wasn't. Compared to the profiles of unknown, vicious, antisocial criminals that I had struggled to compile and make accurate over the past 17 years in the FBI, this attempt at puzzle solving was kid stuff."

JOHN HINCKLEY

When John Hinckley made his assassination attempt on President Ronald Reagan in Washington, D.C., on March 30, 1981, he was immediately apprehended. The FBI quickly established that he was in his mid-20s, unmarried, a college student from Denver, and from a relatively wealthy family. They had the key to his motel room – but they urgently needed more information detailing material in the room so that they could obtain a search warrant.

Robert Ressler was summoned and asked

to provide some indication of what to look for. He reasoned that Hinkley was a loner, not part of an assassination conspiracy, and that his room should be searched for evidence of his loneliness and fantasizing. He named all types of reading materials, such as books and magazines with specific passages underlined, as well as scrapbooks and diaries. In addition, he listed credit cards and receipts, which would help to trace Hinckley's movements over the previous 6 to 12 months; motel bills, which could provide a record of phone calls; and perhaps even a telephone credit card. In particular, Ressler suggested a search for a tape recorder and tapes, which would have been used as a form of diary.

These specific items were named in the application for a search warrant and enabled FBI agents to seize them from the motel room and other rooms Hinckley was known to have used. They found, as predicted, tapes of his telephone conversations with film star Jodie Foster, with whom he had become obsessed. There was a postcard – which had not been sent – with a photograph of the President and his wife, Nancy, on which Hinckley had written:

"Dear Jodie: Don't they make a darling couple? Nancy is downright sexy. One day you and I will occupy the White House and the peasants will drool with envy. Until then, please do your best to remain a virgin…."

There was also a letter addressed to Jodie Foster in which Hinckley said that he was going out to shoot Reagan, and knew he might not return, and he wanted her to know that he had done it for her. There were diaries and comments in newspaper margins and an annotated copy of the script of the movie *Taxi Driver,* in which Jodie Foster had starred. All this material, predicted by Ressler, was sufficient to clinch the prosecution and conviction of John Hinckley.

FBI PROFILE OF JACK THE RIPPER

In October 1988, in a television program entitled "The Secret Identity of Jack the Ripper," which commemorated the centenary of the crimes of the 19th-century killer, John Douglas and Roy Hazelwood were invited to present a modern-day profile of the undetected murderer (the "UNSUB," or "unidentified subject").

Given the choice of all the usual suspects, they agreed that the taunting letters to the police were written as a hoax, as the type of individual who committed the crimes would not be the sort of personality to issue a public challenge of this sort. The mutilations he inflicted

Hinckley in custody after his arrest. It was discovered that his attempt on President Reagan's life was the result of his obsession with film actress Jodie Foster, who was to star, some 10 years later, as an FBI trainee profiler in the movie **The Silence of the Lambs.**

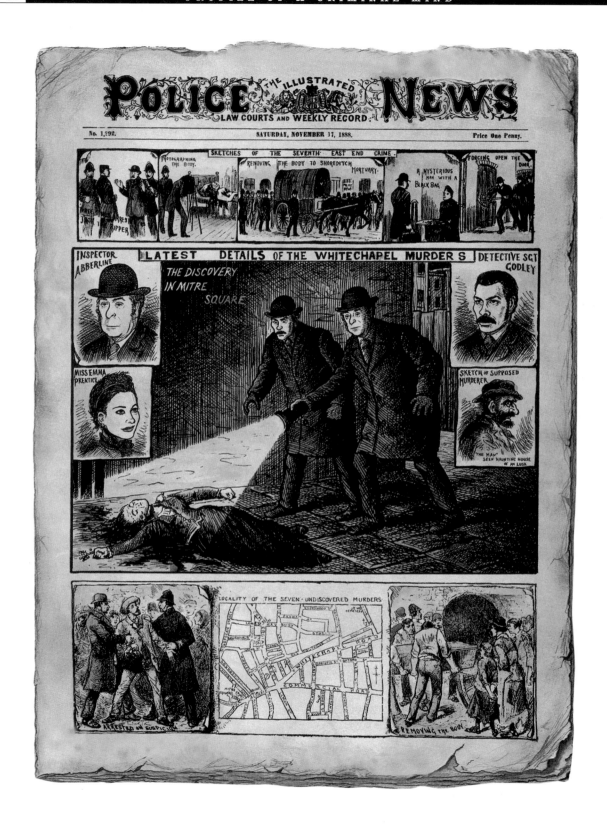

suggested a mentally disturbed, sexually inadequate man, with a generalized rage against women; in removing their organs, he obliterated their sex, so that they were no longer to be feared. The "blitz" style of each attack showed that he was also personally and socially inadequate, and neighbors might well have complained to the police of his behavior.

The killer was a white male in his mid- to late 20s, of reasonably high intelligence, but lucky rather than clever in escaping apprehension. He was single, a loner, who had never been married; he came from a broken home in which a dominant female had physically – and perhaps sexually – abused him. The fact that all the murders were committed between midnight and dawn also suggested that the killer had no domestic commitments.

"Jack" was someone who could blend in with his surroundings, without provoking suspicion or fear. If he were employed, it was in a menial role, which involved little direct contact with other people, and he was certainly not a professional man. In particular, he was among those who had been questioned by the police.

After considering all the possibilities, both FBI agents independently came up with Aaron Kosminski as the likely suspect. Kosminski was a Polish Jew, a boot maker who arrived in London in 1882; he was diagnosed as suffering from syphilis six months before the first of the Ripper murders. It was not until 1959 that a memorandum written in 1894 by Sir

Melville Macnaghten – Asssistant Chief Constable, 1889–1890 – came to public attention. It named Kosminski as one of three known suspects; and in 1987 some marginal notes were discovered in a book written by Sir Robert Anderson, who had early charge of the Ripper investigation.

Anderson wrote: "…the only person who ever had a good view of the murderer unhesitatingly identified the suspect the instant he was confronted with him, but he refused to give evidence against him." The notes in the margin, written by Anderson's predecessor, Chief Inspector Donald Swanson, confirmed that this was Kosminski: "and shortly afterwards the suspect with his hands tied behind his back was sent to Stepney Workhouse and thence to Colney Hatch [a London lunatic asylum]." Kosminski, who was reported to have become "insane owing to many years of indulgence in solitary vices," remained in confinement until his death in 1919. Significantly, no more murders were committed after Kosminski moved to the lunatic asylum.

A portion of the first letter, allegedly from Jack the Ripper, sent to the Central News Agency on September 27, 1888.

Far left: Front page of the weekly Police News, promising the latest details of the investigation into the Whitechapel murders. The main illustration is of the discovery of the Ripper's fourth victim, two days after the above letter was sent.

A SYSTEM OF IDENTIFICATION

During the 1950s and 1960s, an average of 10,000 homicides were committed in the United States each year, and nearly all the cases were solved within 12 months. Most were committed by someone close to the victim: a spouse, a relative, a neighbor, or a colleague, and so were easy to solve. "Stranger-to-stranger" murders were few. However, the situation changed dramatically over the next decade.

In 1980, some 23,000 people were murdered in the United States. Said Lois Haight Herrington, chairperson of the President's Task Force on Victims of Crime:

"Something insidious has happened in America. Crime has made victims of us all. Awareness of its danger affects the way we think, where we live, where we go, what we buy, how we raise our children, and the quality of our lives as we age. The specter of violent crimes and the knowledge that, without warning, any person can be attacked or crippled, robbed, or killed, lurks at the fringes of consciousness. Every citizen of this country is more impoverished, less free, more fearful, and less safe, because of the ever-present threat of the violent criminal."

The year 1980 was also a presidential election year. Republican candidate Ronald Reagan – already known for his tough stand on crime during two terms as

Between 1973 and 1977, Patrick Wayne Kearney (above) and his lover David Douglas Hill murdered 15 young men in California. There was a well-defined signature to every case. All the victims had homosexual backgrounds; each was found nude and shot in the head; and several were dismembered and their remains stuffed into identical plastic bags.

Far right: Henry Lee Lucas was by his own account the most prolific serial killer in the United States – but subsequent investigations revealed that most of his claims were false.

governor of California – devoted much of his campaign to this issue, criticizing Democratic president Jimmy Carter and the U.S. Supreme Court for their failure to deal effectively with the problem. He promised additional powers to law enforcement agencies, and this in part secured his election.

The new attorney general, William French Smith, quickly set up his Task Force on Violent Crime, and enlisted a formidable group of criminologists to explore ways to fight the rising crime wave. He also required each agency of the Department of Justice to prepare a report on how they would be able to assist the national program. Foremost among these was the FBI.

One of the great problems lay in the detection of serial killers. In October 1983, the FBI announced that an estimated 5,000 American citizens had been killed during 1982 by "stranger" murderers, and few of

the cases had been solved. Roger Depue, who had become head of the BSU, stated that there were as many as 35 serial killers at large in the United States; and because their activities could be spread over a wide area, there was no effective means of connecting their crimes and tracing the perpetrators.

Pierce Brooks, retired from the police force as a commander after 35 years, and now a consultant, had been proposing to the Justice Department for some years the establishment of a nationwide, computerized "Violent Criminal Apprehension Program" (VICAP), and in 1982 limited funds were made available by the government to study the proposal. The VICAP Task Force was administered by Brooks from the office of Professor Douglas Moore at Sam Houston State University in Texas, and included homicide investigators, crime analysts, and other experts from more than 20 states and law enforcement agencies, including Robert Ressler from the BSU.

It was during one of their sessions that a task force member hurried into the room and announced that a man named Henry Lee Lucas had just confessed to more than 100 murders in nearly every state in the Union; this, he insisted, was a case that typified the need for VICAP.

HENRY LEE LUCAS

By his own account, Henry Lee Lucas was one of the most horrific serial killers of all time. As a child, Lucas said, he suffered terrible degradation: "I hated all my life. I hated everybody. When I first grew up and I can remember, I was dressed as a girl by Mother. And I stayed that way for three years." He alleged that his mother sent him

Law enforcement officials study the site, near Stoneburg, Texas, where they believed one of Lucas's victims, 80-year-old Kate Rich, was buried. Lucas later led them to a stove near his home, where her bones were found.

a child when a brother stabbed him with a knife. As a teenager, Lucas indulged in all kinds of sadistic and sexually deviant practices. He claimed that, at the early age of 13, he had killed a female schoolteacher because she had rejected his advances. When he was 24, in January 1960, he finally turned on his mother, stabbing her to death. He was found guilty of second-degree murder, sentenced to 40 years, and confined in Ionia State Psychiatric Hospital in Michigan.

Six years later, however, he was released, despite his pleas to the hospital authorities: "I'm not ready to go," he said, "I know I'll kill again." That same day, he claimed, he murdered a young woman within a short distance of the hospital gates. Then, according to his confession, he set off on a murder spree that lasted 17 years, as he roamed all across the United States, raping and killing.

He eventually joined up with arsonist Ottis Elwood Toole and his subnormal, 13-year-old niece, Frieda Powell. This bizarre group spent a year together, on a continuous trail of destruction. When Toole left them Lucas and Frieda settled down together in Stoneburg, Texas, where Lucas got a job as handyman to 80-year-old Kate Rich. When both Mrs. Rich and Frieda disappeared in October 1982, neighbors became suspicious and informed the sheriff. There was no evidence that either of the missing women was dead, and Lucas denied all knowledge of their whereabouts.

In June 1983, Lucas was rearrested on a charge of illegal gun possession, and while in jail he wrote a note for the sheriff: "I have tried to get help for so long and no one will believe me. I have killed for the

to his first day of school wearing a girl's dress, and with a perm in his hair. "After that I was treated like what I call the dog of the family; I was beaten; I was made to do things that no human being would want to do…."

X-rays later taken of Lucas's brain revealed wide damage to the areas that control emotion and behavior, due to his mother's savage beatings. He lost an eye as

past 10 years and no one will believe me. I cannot go on doing this. I also killed the only girl I ever loved."

Then he showed the police where to find the bodies of Mrs. Rich and Frieda, and in the evening began a videotaped confession of his homicidal career. "I've done some pretty bad things," he began. The confession lasted for hours as Lucas admitted what he could remember of over 200 murders in nearly every state.

Following his conviction and sentencing to death, the Texas Rangers received inquiries from police in many states, hoping to clear up a large number of unsolved killings. In the course of these investigations, Lucas was frequently allowed to leave his Texas cell. "He was

conveyed to distant locations by airplane or car, stayed in motels, ate well in restaurants, and was generally treated as a celebrity," wrote Robert Ressler. And police in 35 states decided that they were able to close their files on 210 cases.

The FBI, however, became increasingly doubtful of Lucas's claims. The agent in Houston who interviewed him asked if he had committed the murders in Guyana. "Yep," said Lucas, but he wasn't sure if Guyana was in Louisiana or Texas. In fact, it was thousands of miles away, in Central America, where cult leader Jim Jones had persuaded hundreds of his followers to commit suicide at their Jonestown settlement. In time, records of Lucas's employment, credit card receipts, and other

In jail, Lucas confessed to "way over 200" murders, in nearly every state in the Union. Hoping to clear up hundreds of unsolved cases, police authorities requested his presence for interrogation at many different locations, and announced that they were closing many files as a result.

Here, Henry Lee Lucas stands outside a courthouse, well satisfied with the notoriety his "confessions" had gained. He lost the sight of one eye as a child, when a brother stabbed him with a knife – possibly accidentally.

evidence established that most of his confession was untrue. Much of the investigation was carried out by two journalists on the *Dallas Times Herald* who showed, for example, that Lucas had been in Florida at the time he admitted to a murder in Texas.

When Lucas was finally interviewed by Ressler, he admitted that he had committed almost none of the murders to which he had confessed: since 1975 he had "killed a few," perhaps five. He had told his lies "to have fun," and highlight the "stupidity" of the police. As Ressler wrote: "If we had had VICAP up and running at the time Lucas made his first startling admission, it would have been easy to see what was truth and what was falsehood in his confession."

Lucas spent 15 years on death row, but in June 1999, George W. Bush, then governor of Texas, commuted his sentence to life imprisonment.

THE VICAP PROPOSAL IS ADOPTED

In July 1983, Pierce Brooks and Roger Depue appeared before a Senate subcommittee in Washington, D.C., that was considering the VICAP proposal. In a letter to William H. Webster, the incumbent Director of the FBI, the chairman of the subcommittee, Senator Arlen Specter, wrote:

"Such a system would solicit and analyze information concerning random and senseless murders, attempted murders, and the kidnapping of children.... The

information would be systematized by state and local police agencies, based upon the evidence of one of the specified crimes. This data would be analyzed at a central location and compared with similar assaults. Where a possible connection is identified, two or more agencies would be linked up in their investigations."

Ann Rule, a former police officer and now the author of true-crime books, was among the witnesses appearing before the subcommittee. She said, "The thing that I have found about the serial murderers that I have researched is that they travel constantly; they are 'trollers.' While most of us might put 15,000 to 20,000 miles a year on our cars, several of the serial killers I have researched have put 200,000 miles on their cars. They move constantly. They might drive all night long. They are always looking for the random victim who may cross their path.... The serial killer seldom knows his victims before he seizes them. They are strangers, targets for his tremendous inner rage. He is ruthless, conscienceless, and invariably cunning." This, she emphasized, made a very powerful case for a comprehensive nationwide information program.

Pierce Brooks described the many hours he had spent studying newspapers some 25 years earlier when investigating killings. "Over the years, that primitive system worked for me two or three times," he said. "The real tragedy is that we, the police, are still doing it in the same way today.... What is missing is our ability to analyze crime information and communicate

The 1991 movie *Henry, Portrait of a Serial Killer* was loosely based upon the life and crimes of Henry Lee Lucas.

Senator Arlen Specter, who chaired the 1983 Senate subcommittee that considered the proposal to set up VICAP. In a letter to William H. Webster, director of the FBI, he recommended its adoption.

amongst ourselves, and that is what VICAP is all about." He suggested that the obvious place to set it up was in the BSU at Quantico. Roger Depue then described the work of the BSU and Ressler's research project.

Possibly the most persuasive aspect of the proposal was the estimate that it could cost "as low as half a million dollars" each year. Following a major conference at the Criminal Justice Center of Sam Houston University, William Webster agreed that the FBI concurred in the need for VICAP. He said it should be created by the BSU experts at Quantico, and that the new

program, together with all present BSU functions, should be established separately from the existing National Crime Information Center. He put the cost of this organization at a little under $3 million over the first two years. Nine months later, President Ronald Reagan announced the establishment of the National Center for the Analysis of Violent Crime (NCAVC).

THE VICAP CRIMINAL ANALYSIS REPORT

Up to this point, FBI agents at Quantico had largely relied upon photographs of the scene of the crime to assist them in

providing a detailed description of a crime and its environment. They now drew up a VICAP Crime Analysis Report form, which was distributed to all 59 FBI field divisions.

Designed for computer analysis, the VICAP form was a daunting document at first sight, but every part of it was important. There were 189 questions to be answered, divided under 42 subheads. After noting the necessary administrative details, the FBI agent was required to provide brief information on one of three types of crime: murder or attempted murder; unidentified body, suspected a

homicide victim; or kidnapping or missing person. Also required in this section was an assessment of whether the offender had killed previously and whether the case was related to organized drug trafficking.

Then came the details of the specific case: dates and times; the status (e.g. marital, employment, educational, etc.) and identification of the victim; and a complete physical description of the victim, including birthmarks, tattoos, other outstanding physical features, and clothes. After this, the form required information on people the investigator "has reasonable cause to believe are responsible," as well

William Webster, FBI director, consulting with colleagues. He agreed that the Behavioral Science Unit should introduce VICAP. Nine months later, President Ronald Reagan announced the establishment of the National Center for the Analysis of Violent Crime (NCAVC).

"VICAP's mission is to facilitate cooperation, communication, and coordination between law enforcement agencies and provide support in their efforts to investigate, identify, track, apprehend, and prosecute violent serial offenders."

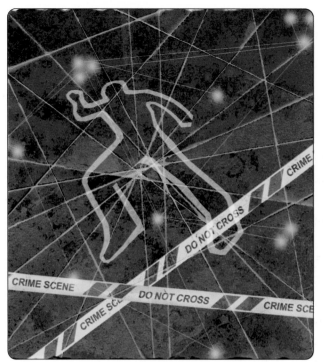

as anyone in custody at present; and details of any vehicle involved in the crime.

So far, 95 questions had been asked. A further 61 questions followed under the heading "Offense MO," covering every detail of the crime scene. Then came 30 questions to be answered by the medical examiner and the forensic investigators. Finally, the agent was required to provide a list of other cases that might be related, as well as a fairly brief "narrative summary," which included "any details of this case you feel are important, but that have not previously been addressed."

Requests for a criminal profile evaluation should not be submitted with the VICAP form. Such a request had to be made to the criminal profile coordinator assigned to the appropriate field division of the FBI. He or she would review the information and, if the request was

approved, submit the entire profile package to the NCAVC.

The FBI issues specific instructions on how agents should fill in the VICAP form. "If in doubt about how to respond to a given item, be guided by your experience and good judgment. Proof beyond a reasonable doubt is not required, but do not guess either.... If your incident has multiple victims, you must complete a separate VICAP Report Form for each victim. Offender information need not be duplicated. If your incident has multiple offenders, submit only one complete VICAP form per victim, Xerox and attach additional offender pages to each report as needed."

Finally, the FBI stressed that: "Cases where the offender has been arrested or identified should be submitted, so unsolved cases in the VICAP system can be linked to known offenders."

In May 1985 Pierce Brooks watched as operators at Quantico fed the details of the first incoming VICAP form into the computer. After nearly 30 years, his dream had come true. In October of that year, the entire cost of the NCAVC was absorbed into the annual budget of the FBI, with four basic programs: VICAP; profiling; research and development (largely involving Ressler's Criminal Personality Research Project); and training of field agents and local police. Ressler, to his great disappointment – "VICAP is a number-cruncher's dream, and I'm more interested in behavioral sciences and in active, ongoing investigations" – was appointed manager of the VICAP program. After some time he retired and became an independent consultant.

VICAP was at first embraced by the majority of the states, and it continues to be employed by the FBI. But many police forces found the full form too complex and time-consuming, and adopted abbreviated reports. The police forces of Rochester, Baltimore, Kansas City, Mobile, Philadelphia, and Chicago; the County of Los Angeles; and the states of New York,

SUMMING UP

The **VICAP** report narrative summary was designed to provide a person unfamiliar with the case to grasp a general idea of what happened. The FBI provides a number of typical examples:

"1: The partially decomposed body of an adult female was discovered in a wooded area of a state park, one quarter-mile from a major state highway. There are indications of sexual assault. Victim died of gunshot wounds. It appears that the victim was not killed at the body recovery site. The victim's whereabouts prior to death have not been established.

"2: Female juvenile was last seen at school. Investigation indicates that she was possibly abducted at or near the school while en route home. The victim has not returned, nor has her body been recovered. Investigation indicates that it is unlikely that the victim is a runaway, or that she disappeared of her own accord. This case is strikingly similar to one that occurred approximately 8 months ago in the same vicinity.

"3: The reported offender entered a locked single-family residence occupied by a man, his wife, and 2 infant children. While the offender was gathering property in the residence, the husband confronted the offender. The husband was shot immediately and died.

The wife responded after hearing the gunshot, and was physically restrained by the offender. The offender hit her repeatedly with his fists, forced her to commit oral sex, and raped her repeatedly. The children were not assaulted. The offender left the residence, and a vehicle was heard to leave the area. Offender arrested during commission of a burglary in the same neighborhood one week later."

When Elizabeth Smart was abducted from her home in Salt Lake City in June 2002, police announced that the FBI had developed a profile of the offender.

Mike Scott, chief of law enforcement at the Texas Department of Public Safety, announces a computer program to link Texas to the VICAP database.

Connecticut, Massachusetts, and Virginia all adopted shorter reports. Other states – including Iowa, Washington, Minnesota, New Jersey, and Pennsylvania – set up their own tracking systems that were not directly linked to the FBI and so lacked communication on a national basis.

SIX STEPS TO ARREST

The FBI profilers have summarized five stages of analysis, leading to the final sixth stage, the arrest of the identified suspect:

1. Profiling input, including the VICAP report. This involves the collection of all available information on the crime, including physical evidence, photographs of the crime scene, autopsy reports and pictures, witness testimony, background of the victim, and police reports.

2. Decision process models. The profiler then organizes the input along several dimensions of criminal activity. In the case of homicide, what is its type (the FBI *Crime Classification Manual* lists 32 major types of homicide)? What was the primary motive – sexual, financial, personal, or emotional disturbance? What level of risk did the victim take before the assault occurred, and what level of risk did the killer take? What was the sequence of acts before and after the killing, and how long did these acts take? Where was the crime committed? Was the body moved, or left where the murder had taken place?

3. Crime assessment. The profiler now attempts to reconstruct the behavior of the offender and his victim. Was the crime organized or disorganized? Was it staged in order to mislead the investigators? What motivations are indicated by such details as time and cause of death, location of wounds, and position of the body? For example: brutal facial injuries suggest the killer knew the victim. Murders committed with whatever weapon happens to be available reflect greater impulsiveness, and suggest that the killer lives near the victim. Murders committed in the morning seldom involve alcohol or drugs.

4. Criminal profile. The typical profile, when completed, includes the offender's race, sex, marital status, residential conditions, and employment history; psychological characteristics, beliefs, and values; probable reactions to the police; and criminal record, including the possibility of similar offenses in the past. Age is also included, but it is the most difficult to pinpoint because emotional development and experience do not always match physical age. At this point, the profiler checks back with stage 2 to ensure that the profile matches the crime data.

5. Investigation. A written report is forwarded to the investigators, who concentrate on suspects who most closely fit the profile. If new evidence emerges at this stage, the profile may be revised.

6. Apprehension. The intended result of this analysis is the arrest of the offender. The key element is then interrogation, which may well lead to a confession, or at least a willingness to discuss the crime.

Since 1990, the FBI's BSU has been renamed Behavioral Science Services, with an Investigative Support Unit handling a Criminal Investigation Analysis Program. In a typical year, the FBI reported that it examined 793 cases, 290 of which fell within FBI jurisdiction. Although summaries of many profiling cases (without names) appear in the FBI's *Crime Classification Manual,* the more specific details of few such cases have been reported – and the majority of these have appeared in the memoirs of retired profilers, such as Robert Ressler and John Douglas. Consequently, it is largely the successes in profiling that have been made public, and not the failures. Little in the way of scientific evaluation of the program has been possible.

Jodie Foster, starring in **The Silence of the Lambs**. *She played the role of Clarice Starling, a trainee FBI agent sent by the BSU to interrogate psychopathic serial murderer Dr. Hannibal Lecter.*

In fact, it has become apparent that, despite the value of the VICAP form in comparing crimes and tracing perpetrators, and despite the instructional programs maintained by the FBI, the work of the individual profilers remains largely intuitive. Nevertheless, an academic study of profiling cases reported in 1990 by A. J. Pinizzotto and N. J. Finkel in the journal *Law & Human Behavior* concluded that profilers can produce more useful and valid criminal profiles than clinical psychologists, or even experienced crime investigators.

THE CANADIAN SYSTEM

In Canada, ViCLAS (Violent Crime Linkage Analysis System) has been instituted by the Royal Canadian Mounted Police (RCMP). This began in the mid-1980s as a computerized databank known originally as the Major Crimes File (MCF). However, by 1990, MCF had some 800 cases on file, no links between reported crimes had been established, and the program was considered a near-failure. As a result, Inspector Ron Mackay, officer in charge of the Violent Crime Analysis Branch of the RCMP at its headquarters in Ottawa, spent 10 months at Quantico, and returned as Canada's first qualified "criminal investigative analyst."

Inspector Mackay concluded that existing American systems needed to be improved to ensure communication across the nation. In particular, he recognized that, because of the distinctions between federal and state law in the United States, the FBI was unable to track serious sexual attacks – which, experience showed, often escalated into homicide. After consulting with a number of psychiatric experts, Mackay and his team put together a report booklet similar to the VICAP form – but even longer, with 262 questions. The Royal Canadian Mounted Police claim that completion of the report takes no more than two hours; "if the investigator can answer every one of the questions, he or she can be assured they have conducted a thorough investigation." In addition to the types of crime considered by the FBI, sexual assaults are also included.

Once research on the ViCLAS questionnaire was completed, a prototype computer system that could assemble and sort the data had to be found. Sgt. Keith Davidson of British Columbia, who had also qualified as a criminal investigative analyst, had developed a local system that he called MaCROS, and this became the preliminary model upon which the Canadian national system was built. Two computer engineers, briefed on police procedures and investigation techniques, set up the necessary software. Initially, each of the 10 ViCLAS units across the country maintained its own provincial database, but now, using modern modems and encryption programs to ensure data security, these are all connected to a central computer in Ottawa. Interpretation of the raw data is in the hands of ViCLAS specialists. They are required to have at least five years'

> After consulting with a number of psychiatric experts, Mackay and his team put together a booklet similar to the VICAP form.

experience of operational police work in the investigation of serious crime, an academic background in the humanities, and a good knowledge of computing.

By May 1997, some 20,000 cases were recorded on the system, and over 3,200 linkages had been established. As these linkages were made, they were put into a "series," and soon over 1,100 series were on file. "These numbers," states the RCMP, "confirm that there are a large number of serial offenders committing crimes against people on a regular basis in Canada."

Dr. David Cavanaugh of Harvard University, who was a consultant on the FBI system, said the Canadians "have done to automated case linkage what the Japanese did with assembly line auto production. They have taken a good American idea and transformed it into the best in the world."

THE DUTCH APPROACH

Within the past 25 years, the Netherlands police have become increasingly aware of the value of crime analysis in their investigations. As a result they set up an Offender Profiling Unit within the National Criminal Intelligence Division of the National Police Agency. Although their

Detective Scott Driemel of the Vancouver Police and RCMP Constable Cate Galliford announce a list of missing women who were feared to be the victims of a serial killer in British Columbia, March 2002.

CRIME ANALYSIS AT INTERPOL

In March 1993, Interpol established a fully operational Analytical Crime Intelligence Unit (ACIU).

They say: "The advantage of crime analysis is that it has introduced structures, methods, and a uniform set of techniques, e.g., an assessment of the scale and nature of high-volume crime such as burglary, numerous varied activities of an organized crime group such as the 'Mafia,' or the identification of a lone serial killer.... Crime analysis is a combination of uniform techniques focusing on the development of hypotheses, reconstructing the course of individual criminal incidents, identifying a series of related crimes, and analyzing the scope of and patterns in criminal activity."

approach was based upon collaboration with the FBI, there were important differences. From the beginning, the unit was made multidisciplinary. For example, a profiling team was likely to comprise both a forensic psychologist and an FBI-trained police officer.

The Dutch also recognized that the unit should be accountable, not only to the police force, but also to the scientific community at large. Members of the unit have been willing to publish the results of their work, opening them up to scientific scrutiny and criticism, as well as exposing them to assessments of the reliability and validity of their claims.

The Dutch system is based on two important principles: that profiling is a combination of investigative experience and a knowledge of behavioral science; and that a profile must be regarded only as a means of indicating the direction that an investigation should take.

Included among the many internal research investigations by the Profiling Unit was a comparison between the inferences that an experienced police detective might make at the crime scene and those that an FBI-trained profiler would find indicative. They found that most detectives' presumptions were subjective, based on personal experience of similar crime investigations, while those officers who followed the FBI's guidelines were more objective and produced more consistent analyses. Peter B. Ainsworth has written about the Dutch approach in *Offender Profiling and Crime Analysis* (2001): "The experienced detective might be more concerned with searching for forensic evidence in the form of fingerprints, DNA samples, fibers, etc.... While the profiler may talk in terms of probabilities or hypotheses, which might be tested, the detective will be more concerned with 'facts.'" Though it is difficult to demonstrate that one approach is better than the other, the profiler's analysis is certainly more scientific.

While the Dutch unit has contributed greatly to the investigation of crime, much of their work is not established "profiling." It was reported in 1997 that "making profiles is now the exception rather than the rule," and attention is now more often concentrated upon the offender's MO.

PROFILING IN FRANCE

In France, most psychological profiling is performed not at the request of the police, but at the request of the examining magistrate, who (not unlike a grand jury) considers all the evidence and decides whether it is sufficient to make a case. He or she can arrange for a clinical psychologist or psychiatrist to examine the complete dossier – police reports, witness statements, photos and plans of the crime scene (the profiler rarely visits the site), autopsy photographs and reports, and forensic details – and make an assessment on this basis. Great emphasis is also placed upon what can be discovered of the psychology of the victim.

Clinical psychologist Pierre Leclair is said to be the sole exception to this legal practice. After working for 15 years in a psychiatric hospital, he joined the French national police to advise on recruitment, but has devoted the past few years to offender profiling.

However, Leclair does not wish to be known as a "profiler." "It is a term that cannot be applied in France," he stated. "When there is a crime in South Africa, my colleague there will have a helicopter land in his garden to carry him to the other end

Police officers from many different countries in a conference at Interpol's Paris headquarters.

of the country. We are far from that in France." He requires a week "to imbibe the dossier," and works closely with the investigators: "I am not there to do the work of the police, but to assist them." He covered 32 criminal cases over the course of three years, and claims that the best of his profiles were "60 percent useful."

Little has been published on psychological profiling in Germany or Switzerland, although it is known that the police in these and other countries are interested in the technique. The situation is very different in Britain, where a succession of mass-market books and a successful TV series have attracted wide popular attention.

The French serial killer, Francis Heaulme, a vagrant. He has been accused of the motiveless deaths of more than 50 men, women, and children – although he denies that he has ever killed a child.

THE DEVIL COP

In the early 1990s, a French gendarme who was dissatisfied with the support he was given in his three-year-long pursuit of a serial killer resigned and set himself up as a private detective specializing in psychological profiling. This man was Jean-François Abgrall, who was described as "that devil" by defense lawyers when he successfully obtained the conviction of Francis Heaulme in 1994.

The two men first met in a gendarmerie (police barracks) in Normandy in 1989, when Abgrall was investigating the murder of a woman in Brest in nearby Brittany. Heaulme had an apparently watertight alibi: at the time of the murder, a nurse in a hospital 80 km (50 miles) away, where he was being treated for alcoholism and the debilitating effects of his vagrant lifestyle, had noted his temperature.

"As soon as I saw this man," Abgrall wrote in his book, *Dans la Tête du Tueur* (*In the Mind of the Killer*, 2002), "I detected a dangerous and violent man.... The watertight alibi? I discovered that, if a patient was absent, nurses noted the temperature from the thermometer on the bedside table."

The suspect was a wandering misfit, an alcoholic who spent many brief periods in hospitals. Abgrall doggedly tracked his erratic movements and spent many hours interviewing him, gradually establishing an understanding of the man's personality. Heaulme spoke in a strangely abstracted way about *pépins* ("obstacles"), which had occurred on specific dates between 1986 and 1991, when he claimed to have been a witness to murders committed by other people. Abgrall wrote: "We realized that he was talking about crimes we did not even know had been committed, and we let him lead us into his world at his own pace."

There was no consistent pattern to the murders, and the authorities were unable to place Heaulme at the scene of the crimes. Despite his odd and evasive behavior, a psychological assessment reported: "Heaulme cannot be considered dangerous in the psychiatric sense; on the other hand, it appears that he is dangerous in the criminological sense." Sufficient evidence was eventually gathered to show that Heaulme was guilty of the Brest murder, and the French police believe that he was responsible for more than 50 other murders of men, women, and children.

Jean-François Abgrall, a former French *gendarme* who resigned and set up as a private psychological profiler. It was his persistent pursuit of Francis Heaulme that led to the serial killer's apprehension.

THE INTUITIVE APPROACH TO CRIME ANALYSIS

The British Isles have had their share of serial murderers, few more horrific than husband-and-wife team Fred and Rosemary West (left). Profiling in Britain has developed in different directions and remains primarily an intuitive interpretation, not done by computers or criminal investigators but by criminal psychologists.

The adoption of criminal profiling in Britain came about in an almost accidental way. David Canter was Professor of Applied Psychology at the University of Surrey when he was invited to lunch at London's Scotland Yard in November 1985. There he discussed with two top detectives whether behavioral science could contribute to police investigations.

"At that time I had never heard of 'profiling,' but the whole idea of reading a criminal's life from the details of how he carries out his crime was enormously appealing," he wrote in his book *Criminal Shadows* (1994).

Two months later, Canter read a front-page feature in London's *Evening Standard* that described a series of 24 sexual assaults over the past four years; the article stated that investigators believed all were by the same assailant – although sometimes he had an accomplice. Using the newspaper report as his source, Canter drew up a chart of events and the dates when they had occurred, separating those that involved two assailants from the single assaults.

He came to the realization that the cases when the individual acted alone had taken place relatively recently, while both attackers had been involved in the earlier assaults. In a letter to Scotland Yard, he wrote: "I have no evidence that the one

individual is the same person, but if it were then one could see something about the relationship between the two men as a possible clue to the whole series. For example, a scenario that had one of them working near the railways, mentioned in the news, and only meeting up with the other under certain circumstances, possibly related to work, could lead to exploring the evidence...."

CATCHING THE "RAILWAY RAPIST"

To his considerable surprise, Canter was invited some months later to a high-level conference at Hendon Police College in north London to discuss the investigation. While there, he learned that three police authorities were involved in the case because two separate murders – in addition to the sexual assaults – were also being investigated.

Forensic evidence and certain unusual aspects of the crime scene led the police to conclude that the rapes and the murders had all been committed by the same man, dubbed the "Railway Rapist" by the press. To Canter's even greater surprise, he was asked if he would assist the police in their investigation. Two officers were assigned to help him in his work.

The two murders were of a 19-year-old secretary whose body had been found in East London and of a 15-year-old Dutch girl (recently arrived in England with her family) whose body was discovered in a forest 40 miles south of London. The police were on the verge of closing down the inquiry into the series of rapes, due to lack of progress, when the link with the murders was established.

Both rape and murder victims had their thumbs tied together with a particular type of twine and the offender had set fire to tissues with which he had wiped the victims. A specific blood group had also been found in the rape cases and on the body of the Dutch girl.

Canter was given a small research room at police headquarters in Guildford, near the university where he taught, and there he was joined by the two police officers. He brought in a desktop computer – and a software program that had been paid for out of fees he had received for doing market research on people's cookie

Far left: It was known that John Duffy (left), during his earlier series of rapes, had committed them with another man. When Professor David Canter's analysis led police to Duffy, his accomplice was soon identified as David Mulcahy, a friend from childhood who had assisted him in his hunt for victims. The two are seen in a photo-booth photograph taken during their teenage years.

THE CRUCIAL MAPS

In the course of their analysis, Canter and his team decided to sketch maps of the rape and murder sites for each year in which they occurred. These were drawn on separate acetate overlays.

As Canter looked at the overlays for the three earliest incidents, in 1982, he said, with a questioning smile, "He lives there, doesn't he?" All three incidents took place close together, around the Kilburn district of North London.

As FBI profilers had independently determined, most serial killers and rapists begin committing crimes close to where they live, gradually extending their activities further afield as they gain in experience and confidence. Each of the yearly maps clearly demonstrated this.

preferences. Together Canter and the policemen entered every detail that was known about each crime to search for patterns that might reveal that the same man was responsible. They were analyzing his behavioral characteristics because the physical descriptions provided by the victims under intense emotional stress were extremely inconsistent.

After the team had been working for only a few weeks a third murder occurred, which bore all the hallmarks of the previous two. A 30-year-old secretary had disappeared after taking a train to her home north of London, and her body was discovered two months later. It was badly decomposed, but investigators still noted that her fingers were characteristically tied together and an attempt had been made to burn the body. It was now becoming increasingly urgent to find the killer before he struck again.

MAPPING THE CRIMES
By July 1986, Canter was under pressure from the investigating team to come up with a preliminary profile. He announced that the killer had lived in the Kilburn area, or close by, since at least 1983, probably with a wife or girlfriend, but without children. He had a semiskilled or skilled job involving weekend or casual labor that did not usually bring him into contact with the general public. He also had knowledge of the railroad system of the London suburbs. Although he kept to himself and had very little to do with women he had one or two very close male friends.

Far right: John Francis Duffy, the Railway Rapist. Suspected of three murders of young girls, he was found guilty of two, together with four rapes. His accomplice, David Mulcahy, tried subsequently, admitted to 22 attacks on women over 12 years.

Taking the most consistent of the victims' descriptions and some of the forensic findings, Canter concluded that the offender was in his mid- to late 20s, below average height, and had light hair. He was right-handed and of blood group A. Finally, Canter concluded that the man had been arrested at some time between October 24, 1982 and January 1984 (there had been only one attack in 1983). The arrest might not have been for a sex-related crime, but an aggressive attack, possibly under the influence of alcohol or drugs.

The police assembled a list of nearly 2,000 suspects; Duffy's name had been included because he had been arrested in 1985.

Four months had passed when Canter received a telephone call from the senior police officer in charge of the investigation, who said, "I don't know how you did it…but that profile you gave us was very accurate…." John Francis Duffy had been arrested and charged with the crimes. The police had assembled a list of nearly 2,000 suspects; Duffy's name had been included because he had been arrested and released on bail in July 1985 for an attack on his estranged wife – but he was number 1,505 on the list. His description fit that of a man who had raped a woman in September of that year, but his victim failed to identify him in a police line-up.

Duffy was picked up again by the police near an isolated railroad station in May 1986. He was found to be in possession of a knife and a wad of paper tissues, but even this evidence was considered insufficient to charge him. It was only following Canter's profile that Duffy rose to the top of the suspects list. At his trial,

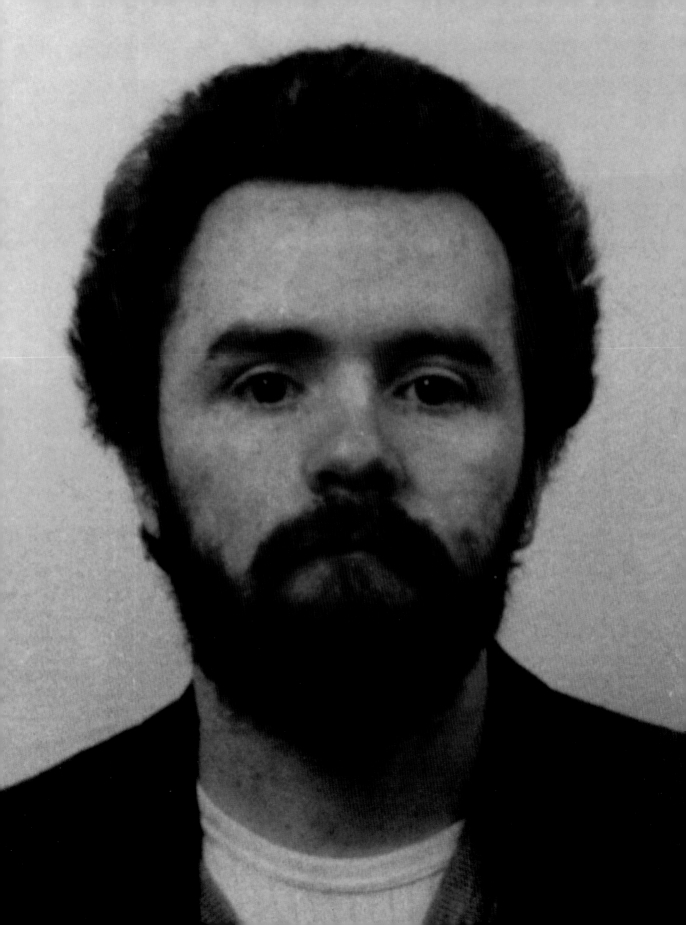

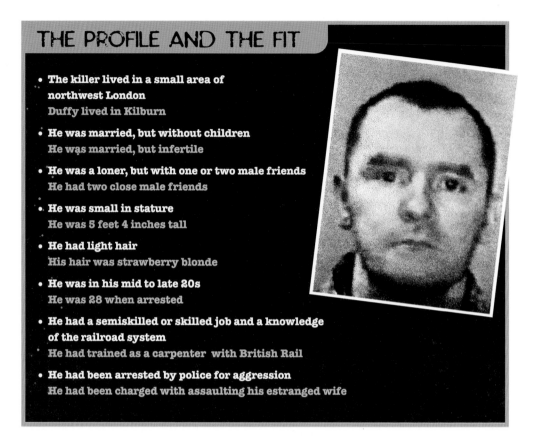

THE PROFILE AND THE FIT

- **The killer lived in a small area of northwest London**
  Duffy lived in Kilburn

- **He was married, but without children**
  He was married, but infertile

- **He was a loner, but with one or two male friends**
  He had two close male friends

- **He was small in stature**
  He was 5 feet 4 inches tall

- **He had light hair**
  His hair was strawberry blonde

- **He was in his mid to late 20s**
  He was 28 when arrested

- **He had a semiskilled or skilled job and a knowledge of the railroad system**
  He had trained as a carpenter with British Rail

- **He had been arrested by police for aggression**
  He had been charged with assaulting his estranged wife

Far right: Professor David Canter made his first psychological analysis of a criminal in the case of John Duffy. His profile of the offender helped police identify Duffy among their long list of suspects. Canter published a description of his methods, which he named "investigative psychology," in **Criminal Shadows** *in 1994.*

he was found guilty of four rapes and two murders – the evidence in the third being considered insufficient by the prosecution – and sentenced to 30 years in prison.

Forensic evidence finally clinched the identification. At Duffy's mother's house, police found a ball of the unusual twine he had used to tie his victims' hands (it also emerged that Duffy had enjoyed tying up his wife for sex). In addition, 13 "foreign" fibers discovered on the clothes of one of the murdered girls matched those from one of his sweaters.

Canter's contribution to the Duffy case drew enormous publicity, which led to his involvement in a number of subsequent cases. One of these came in 1988 following a series of sexual assaults in Birmingham.

THE BIRMINGHAM RAPIST

Between January 1986 and March 1988, seven elderly women in high-rise apartment buildings in two areas of Birmingham were assaulted. Women in their 70s and 80s were followed into the elevators by a young, athletic, black man. He took them to the top floor, sometimes carrying them the last two flights, where he sexually assaulted them. Although he was physically strong, he exhibited an unexpected gentleness. In a few cases, when the victims complained about the cold floor on which he made them lie naked, he placed some of their clothing underneath them. The man's behavior and choice of victims puzzled the police. Their problems were compounded by the difficulty of making sense of the traumatized victims' accounts.

Canter was struck by the offender's limited choice of locations and victims. Following the man's conviction, Canter took a helicopter flight over the high-rise apartments on the edge of the city center. Only then did he fully realize what a significant feature they were in the landscape. Surrounded and separated by busy road systems, each stood out as a secluded island. By studying detailed maps, Canter had already come to the conclusion that the offender lived not far from the sites of his attacks, probably in a similar high-rise, and knew how to move from one "island" to another. With so many inhabitants, the apartments ("streets in the sky," as Canter described them) were visited every day by many people – most of them unknown or unfamiliar, but attracting little attention.

Apart from these important considerations, Canter had to determine what sort of criminal was involved. "Was the offender a person used to going into other people's private realms, like an experienced burglar would be? Or was he an impulsive attacker, like someone who would grab at a victim in a public place because the opportunity is seen to present itself?...Theft or burglary appeared to be no part

of his actions. Here was a man more like those we would expect to attack outside, seeking the opportunity to find a victim. The streets in the sky were just that."

High-rise apartment blocks in the West Midlands city of Birmingham, which David Canter has aptly described as "streets in the sky." Like urban thoroughfares, they are busy with people coming and going and provide suitable anonymity for a prowling rapist.

The man made no attempt to disguise himself, indicating that he was unlikely to be an inhabitant of any of the apartments in which the assaults took place. But he was clearly familiar with the elevators, passageways, and service doors typical of these buildings.

In a report to the police that covered these matters in detail, Canter and a colleague included a personality assessment. They concluded that the offender was immature sexually and unlikely to have had relationships with a female of his own age group. It was possible that he had dealings with elderly people "in a non-offense context" (for example, he might work as a caregiver). His reported cleanliness and hairstyle and his consistent choice of setting suggested an obsessive personality. Changes in the timing of his attacks and in his style of dress suggested that he could have moved from schooling to employment around October 1986. Finally, they listed four likely areas of residence in decreasing order of probability.

These suggestions were circulated to all the police forces in the Midlands region. When a report came in of a further assault, the senior police officer on duty in Birmingham that day wondered if the offender might have a previous conviction for a minor sexual offense. Going through the records in alphabetical order, he soon came across the name of Adrian Babb, who had been found guilty of attempting to put his hand between the legs of a 60-year-old woman in Birmingham Central Library. Babb lived in the area named as most likely in Canter's report, and investigation of the scene of the latest assault revealed a fingerprint, which was quickly identified as his.

IT WAS BECAUSE OF THEIR AGE

When arrested, Adrian Babb cooperated fully with his police interrogators, who used the profiling report to test the logic and psychological validity of what he told them. He claimed that his attacks were due to his having failed in relationships with a girl and a married woman, but both when interviewed denied that there had been any close involvement. Asked why he had picked on elderly women he replied, "Their age," although he also implied that it was retaliation for "getting done for the one in the library, could be sort of like revenge, don't know."

As an only child, Babb had been friendly with a family of four sisters and had regularly stolen their underwear. Police found the collection at his home, neatly labeled with the owners' names. He had a job as a swimming-pool attendant – which explained his cleanliness and the times of his assaults – but the only friends he recalled from the pool were nine- and ten-year-old girls, a sure sign of his sexual immaturity. After a legal argument in court, Babb was judged sane, and he was sentenced to 16 years' imprisonment. The police told Canter: "Your report was so accurate we thought you had his name and address but were keeping it from us."

THE CAMPUS ATTACKS

Another case of serial rape in the same city soon followed this first case. All 10 victims were female students at Birmingham University, and in every case the victim was asleep when the assailant entered her lodgings. The police were convinced that one man was responsible.

By this time, Canter and his colleagues had completed an analysis of some 60 solved rape cases and found that the actions of the offender could be broken down into discrete behavioral components. They isolated 10 specific aspects of behavior, such as the way in which the offender controlled his victim, his conversation, and his sexual actions. They then used a computer program that draws a square plot and represents each crime as a dot within that square to chart the behavior: the closer the individual dots, the more similar the pattern of behavior.

The first computer "maps" generated by the computer caused considerable concern. No matter how the analysis was carried out, three crimes fell on the left-hand side of the plot, while the other seven were on the right. The three on the left had involved threats of violence, but in the others the offender reassured the victim that she would not be hurt. Either the offender exhibited markedly varied behavior or had changed his style of criminal behavior, or more than one man was involved.

THE "WIMP" AND THE "MACHO MAN"

In April 1989, just two weeks after they had been consulted and because they were concerned that the rapes might soon escalate into murder, Canter's team sent their first report to the police. They concluded that the man had planned his assaults and drew on his experience of breaking into houses. Previous burglaries in some of the premises could well have been committed by the same person. Their report suggested that this person might have a criminal history and a consequent knowledge of police procedure in inquiries. The consistent late-night occurrence of the attacks suggested that the offender was returning from a night shift or on his way to early-morning work. Perhaps doubting the validity of their computer plot, the team did not at that time mention the possibility that there might be two separate offenders.

However, the police had obtained samples for DNA analysis from three of the crimes and when the results came through in late May they were justifiably annoyed: the DNA from rape number 10 did not match that from numbers 4 and 9 – there were obviously two offenders. Their suspicion that there were two rapists confirmed, Canter's team returned to their analysis of the offenders, whom they nicknamed "the wimp" and "macho man."

They decided that the "wimp" was not an experienced burglar. In fact, it was likely that he had no previous criminal history. He was a lonely man who did not have regular contact with women. Older than the typical rapist, he probably wandered the streets, peeping into windows. The sites of his offenses indicated that he lived in the southeastern part of Birmingham. "Macho man," on the other hand, revealed many characteristics of an experienced criminal and was used to exerting control over women. He probably lived in the northwestern area of the city. The police had assembled a list of suspects and eventually

The epithelial cells of the interior of the mouth are constantly detaching, and they provide the easiest way of obtaining a sample for DNA analysis. The cells can be gathered by the use of a simple swab such as this one. DNA testing helped the police establish that there were two different offenders assaulting women in the Birmingham University campus attacks.

DNA analysis identified two of them.

The "wimp" was a college failure, a 27-year-old "mother's boy" who had previously been questioned for prowling and peeping. "Macho man" was 23 years old. He already had an extensive criminal record and was suspected of dealing drugs and pimping. In their police photographs, the "wimp" appears puzzled and bespectacled; "macho man" stares confidently into the camera.

MAPPING THE MIND OF A RAPIST

Canter and his colleagues soon realized that the cases brought to them by the police were not typical: they were cases that the police were finding difficult to solve by their usual procedures, and many significant details did not emerge until the offenders were apprehended. They therefore continued their analysis of crimes that had been solved. They began with rape, and subsequently moved on to study cases of murder, the sexual abuse of children, fraud, and extortion.

Their behavioral analysis of rape cases resulted in a valuable computer plot. At the center of the "map" lie those actions characteristic of rape in general: surprise attack, removal of clothing, and the physical act. Surrounding this area are three segments that reveal how the rapist views his victim. If the rapist sees his victim as an "object," he will have made preparations, perhaps disguising himself, carrying a "rape kit," or using a weapon to exert control. If he sees his victim as a "vehicle," he will be openly aggressive, insulting or demeaning the victim, and enforcing a variety of sexual activities that require the victim's participation.

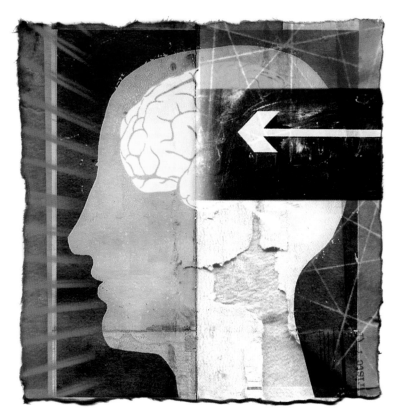

If the rapist sees his victim as a real "person," he will attempt to establish greater social intimacy, perhaps starting a conversation and even complimenting her on her appearance and asking about her life and interests.

Canter believes that these three segments represent the distinct ways in which the offender regards himself. The first is an organized criminal who will exert control to get what he wants. The second is violently aggressive, venting his anger on the victim. The third is fundamentally insecure and carries out his rapes in the hope of establishing some sort of emotional relationship.

David Canter continues his work, which he calls "investigative psychology," and is frequently called on by the police.

David Canter and his coworkers have developed a computer program that plots the behavioral characteristics of a rapist on a two-dimensional "map." They can use this map to identify three distinct ways in which the criminal regards himself.

A DIFFERENT APPROACH

At the same time David Canter was first developing his computer-based behavioral analysis of a serial offender, another British psychologist was taking a very different approach to a similar problem – his analysis was largely intuitive. He was Paul Britton, a clinical psychologist in Leicestershire, whose first book, *The Jigsaw Man,* was published in 1997. Like Canter, his first involvement with a police murder investigation had come about almost by accident, after he had been consulted in a case of a young woman infatuated with a police officer.

In 1984, as a result of this case, he was invited to give his opinion concerning the unsolved murder of a 33-year-old woman on a local canal towpath, in July of the previous year. The officer in charge of the investigation asked him: "If I were to show you the scene of a crime…is it possible for you to tell me things about the person who was responsible for the murder?" Although doubtful, Britton agreed to try.

Photographs showed that the victim's hands and feet had been bound with twine, and she had been stabbed five times in the neck and twice in the chest. There were no signs of robbery or sexual assault. A piece of paper, found near the body, included a sketch of a pentagram in a circle – a symbol often associated with black magic rituals. The police had already interviewed more than 15,000 people, and 80 men had been arrested on suspicion before being released.

After brooding over the case for three days, Britton summarized his conclusions. He decided that the pentagram did not signify that the murder was a ritual killing, but that it represented a rationalization by the murderer for his sexually deviant urges. The random nature of the stabs suggested a young man in his mid-teens to early 20s. He was likely to be lonely and sexually immature with few previous girlfriends, if any, because he lacked the necessary social skills. He probably lived at home with his parents near the scene of the crime and knew his victim by sight.

The killer was likely to be a manual worker, experienced with sharp knives. The ease with which he had overcome the victim and the force behind the stabs indicated that he was strong and athletic. "His violent sexual fantasies," wrote Britton, "will be fed by pornographic magazines…some violent and featuring satanic themes. When you find him…I expect you'll find ample evidence of this, as well as his strong interest in knives."

BEHAVIORAL PATTERNS

As David Canter wrote in *Criminal Shadows*:

"Here, more than anywhere else, the difference between behavioral science and Sherlock Holmes and beyond is most clear. The comparison of behavioral profiles is a comparison of patterns, not the linking of one clue to one inference. The term 'offender profiling' does appropriately draw attention to the configuration of the many points that a profile must have. One point, or clue, no matter how dramatic, does not make a profile."

MAKING A BEHAVIORAL "MAP"

To make a behavioral map, each crime is entered on a computer, one below the other in a row. The types of behavior are entered as columns. For each crime, an action that did not occur is entered in the appropriate column as 1, one that did occur as 2. The resultant chart is called a "data matrix." In rows where the pattern of 1s and 2s is very similar, there is a likely correlation between the crimes.

The program used by David Canter represented each row of numbers – that is, each crime – as a point located somewhere within a square. The computer connects all of these points to each other and then tries to relate them to each other as closely as possible. The more similar the behavior in two crimes is, the closer the points representing those crimes will be.

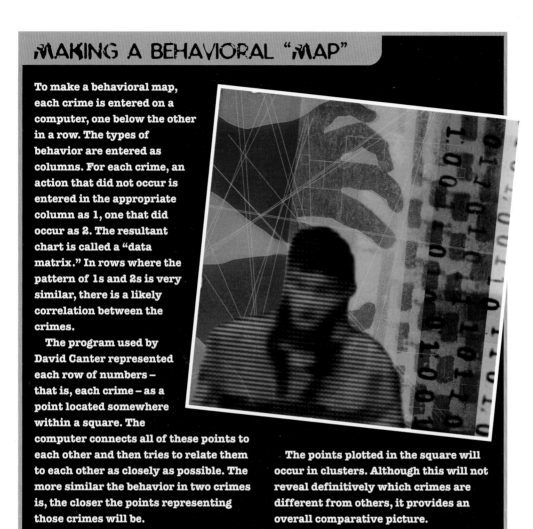

The points plotted in the square will occur in clusters. Although this will not reveal definitively which crimes are different from others, it provides an overall comparative picture.

THE "PENTAGRAM KILLER" STRIKES AGAIN

Fourteen months passed before a similar murder occurred. A 21-year-old nurse on a pedestrian walkway was killed, and although she had not been tied, the pattern of knife wounds was the same. Summoned to police headquarters, Britton agreed that the killing was by the same man. But the attack had been sudden, he said, and it seemed unlikely that the man knew his victim. Something about her had attracted his attention when he was most sexually aroused. Witnesses were able to describe a man seen in the vicinity at the critical time.

One suspect who fit the description was Paul Bostock, a 19-year-old meat processor. He also matched Britton's assessment. Searching his bedroom in the house where he lived with his parents, police found a collection of knives, martial arts weapons, pornographic magazines, and crude drawings of women being tortured. However, he gave nothing away during interrogation.

Britton advised the police to be very circumspect in their questioning, regularly withdrawing from the central facts of the murders, but at the same time letting Bostock know that they understood the problems of sexual deviancy. Eventually Bostock confessed everything. And why had he selected the nurse? "Because she wore red shoes," he said.

Paul Britton, the British clinical psychologist whose approach is largely intuitive. Britton had considerable success in advising police in a wide variety of cases, including that of Fred and Rosemary West, the Gloucester couple that was responsible for the deaths of at least nine young people, including their own daughter.

When Paul Bostock murdered his second victim, the pattern of knife wounds was the same, but the attack seemed spontaneous. Britton suggested that something about the young woman had attracted Bostock. "She wore red shoes," said Bostock when he confessed.

CASE STUDY: THE PET AND BABY FOOD POISONER

In 1988, Britton was presented with a very different kind of crime: blackmail. The managing director of a company that manufactured pet foods had received a can of the company's dog food in the mail, accompanied by a typewritten letter. The food in the can had been contaminated (as the letter stated) with "chemicals... colorless, odorless, and highly toxic." The writer threatened to place similar cans on store shelves throughout Britain and demanded a total of £500,000 (about $800,000) to be paid in installments into various accounts.

Britton was asked to join the Leicester police team dealing with the threat. He assessed the blackmailer as being of average or above average intelligence, but probably not possessing a university education. He was almost certainly working alone, showed persistence, and planned things carefully. This suggested that he was not a young man, but fully mature and very patient.

The man obviously knew how the police would attempt to track him down: he seemed to be aware of unpublished details of a previous similar case. He made all his arrangements by mail using an alias and collected his mail from a private address box in West London. Britton advised the company to begin paying limited amounts into one of the named bank accounts. The man would only be able to withdraw money from automatic teller machines, and his identification number would enable the police to track his movements. At the same time, Leicestershire Criminal

The Heinz baby food factory in Wigan, Lancashire. For a long time, the company refused to pay money to the blackmailer contaminating its jars of food, but eventually Britton persuaded the people at Heinz to deposit relatively small sums into named accounts. In this way, the police were eventually able to apprehend the blackmailer.

Investigation Department (CID) enlisted the assistance of detectives all over the country to mount undercover surveillance on hundreds of cash machines.

However, the blackmailer began to make almost daily withdrawals in towns and cities hundreds of miles apart and generally late at night. Looking at a map of the progressive withdrawals, Britton noticed that they occurred on the highway system that radiated from London, where a significant number of withdrawals were also being made. It seemed likely that the blackmailer lived in or near London, and since he was free to travel long distances overnight, he was possibly retired and almost certainly living alone. The pattern of withdrawals in the vicinity of London also suggested that his home might be to the east of the city, in the area of Hornchurch, in Essex.

Like the "mad bomber," the blackmailer became enraged by the pet food company's protracted negotiations. He began to place contaminated cans on supermarket shelves – accompanied by a telephoned warning – and demanded even more money. By March 1989, the sum had risen to £1,250,000 (over $2,000,000) and the number of detected contaminations rose to 14. Britton advised the company: "If you don't withdraw the product, and you don't go public, he'll think he still has control, and will push for the money."

THE STORY GOES PUBLIC

A national newspaper got wind of the case and published details. Within days, the blackmailer turned his attention to baby foods manufactured by Heinz, with a demand for £300,000 (around $500,000) and the threat that no warnings

would be given. The case now became the concern of New Scotland Yard, where Britton attended a meeting. At the end he shyly ventured the opinion that: "…he is, or was, a police officer…. I'd say he's possibly retired, on suspension, or on sick leave; however, he seems to know so much about what's going on, you have to look at someone who has an inside contact." The police were scandalized by the suggestion.

Over the following months, hundreds of jars of contaminated baby food – some containing caustic soda, others broken razor blades – were reported. The police thought that many were copycat cases but there was no doubt that a significant proportion was attributable to the blackmailer – and his demands rose to £1.2 million (nearly $2 million). Heinz deposited £19,000 (more than $30,000) into two nominated accounts and cash was regularly withdrawn. Despite instituting a secret internal inquiry following Britton's suggestion, police seemed no closer to apprehending the blackmailer.

Then, on the night of October 20, they got lucky. A surveillance team saw a man – inexplicably carrying a crash helmet – leave his car and approach a cash machine. When challenged, he said, "No problem, guys, I know what this is about, but I am innocent," and fainted.

The man turned out to be a 43-year-old former police detective, Rodney Whitchelo. His wallet contained the relevant cash machine cards, and a search of his home revealed materials with which he had contaminated the foods. And he lived in Hornchurch, in Essex. He had regularly mixed with former colleagues and even dropped in on the Crime Squad office to discover how the investigation was going. On at least one occasion, he had sat with police friends in their surveillance car!

Rodney Whitchelo, the blackmailer, leaves his home on the eastern outskirts of London.

The police were highly impressed with Britton's assessment. He later wrote, "I'd never seen his photograph or heard the sound of his voice, but I knew what went on inside his head."

CASE STUDY: MICHAEL SAMS

On July 9, 1991, 18-year-old Julie Dart disappeared in Leeds, Yorkshire. Two days later, two letters were mailed from Huntingdon, Cambridgeshire, about 150 miles away. One went to Julie's boyfriend and was apparently in her handwriting: it reported that she had been kidnapped and that the police should be informed immediately. The other, addressed to "Leeds City Police" (a department that no longer existed, having been absorbed into West Yorkshire Police), was typewritten. It said that a young prostitute had been kidnapped in the red light district of Leeds and demanded that two cash payments – of £140,000 (around $230,000) and £5,000 (about $8,000) – be deposited into two bank accounts. If the details set out in the letter were not followed exactly the girl would be killed and a firebomb set in a major city store somewhere in Britain.

The details were explicit and typical of many modern ransom demands. A female police constable was to await a telephone call at a specific phone booth at Birmingham New Street railroad station, where she would be given further instructions about where the money was to be taken. On July 16, a female officer was positioned by the specified telephone.

Far right: Paul Dart carries a photograph of his sister Julie as he leaves church with their mother, Lynn, following the funeral service for the murdered girl.

The details were explicit and typical: a female police constable was to await a telephone call at a specific phone booth.

It rang, but there was no response from the other end. Four days later, Julie's body was found in a field in Lincolnshire. It was estimated that she had been dead for as many as ten days – probably before the letters had been mailed.

There was no evidence of sexual assault: Julie had been made unconscious by two precise blows with a blunt instrument to the back of the head, and then strangled. Three days later, the police received another letter. The writer expressed his regret at Julie's death, implying (falsely) that it was the result of the failed telephone call, repeated his threats, and gave instructions for another contact by phone. Detective Chief Superintendent Bob Taylor decided to consult Paul Britton.

THE GAME PLAYER

Britton's opinion was that the kidnapper's principal motivation was not money but a desire to play games with the police. "His letters read like shopping lists of demands, specifying how the money should be wrapped, the thickness of the plastic, and exact dimensions of the bundle. Why be so specific? Does it matter? Only if you want to exert control, making your opponent jump through as many hoops as possible.… The blackmailer knew he was clever and now he wanted the police to show him some respect.… The letters contained no evidence of loss of control, or anger, but clear signs of a quiet enjoyment. It must have taken him a long while to plan it all – every hour enjoyable."

At a top-level meeting in Leeds, Britton detailed the profile he had drawn up. The

A composite sketch of the wanted man issued by police. Stephanie Slater was able to give a detailed description of her kidnapper.

school and was familiar with electronics and machinery, principally at a theoretical level; however, the style of the letters indicated that he was not a senior employee in a large organization.

He had probably been married but was unable to sustain such a relationship over any length of time. "He's likely to have a previous history of offending," said Britton, "but this will be for things like property offenses, deception, fraud, and misrepresentation… he clearly has some knowledge of police procedure…. He's almost certainly creating blind alleys for you, and will be hiding behind a welter of smokescreens and diversions." He concluded that the man had planned all along to kill Julie Dart because he wanted to be taken seriously by the police.

Asked where the offender might be based, Britton pointed out the large triangle formed by Leeds, Birmingham, and Huntingdon. "He has strong geographical connections with the West Midlands," he said, "and I think he'll live somewhere inside the triangle. If you want me to put my neck out, I'd favor the right-hand corner, closer to where Julie was found." On July 30, the police received an extraordinary letter, mailed from Leeds. In it, the writer discussed ("after 2 hours pondering this on the train") the risks to himself in 14 different

offender was in his late 40s to early 50s – his methodology had a significant maturity about it. He was of above average intelligence, but not educated at a university. It was likely that he had attended further courses after leaving

scenarios, and detailed the "game odds" of winning or losing. It seemed that Britton's assessment had been very near the mark.

The letter also asserted that another prostitute would soon be abducted. Once again, the ransom instructions were followed on August 6; again, there was no telephone contact.

On August 8, another typewritten letter was received by the police, with different instructions, and on August 14 a female officer at the stipulated phone booth heard a man speaking on the telephone. He told her that he had abducted a prostitute named Sarah Davis in Ipswich, in the county of Suffolk.

A TRAIL OF ENVELOPES

The police in Ipswich were immediately contacted, but they had no reports of any missing prostitutes. However, the next morning a strange discovery was made below a bridge across a former railroad track in Yorkshire, now a ramblers' path: a brown envelope attached to a white-painted brick and beside it a small, silver-painted container with two red lights on top with a coil of wire protruding from it.

Fearful that the device might be a bomb, the police called in an Army disposal team. The container was destroyed, but the envelope survived and in it was a message – numbered 3 – directing the finder to a footbridge over the nearby highway. Here there was another white-painted brick, but no message. However, the police deduced that this was part of a chain intended to lead the carrier of the ransom money from one spot to another.

It was typical of the type of ransom demand made by a kidnapper who wants to be sure that the carrier is alone; he watches from a distance. The police backtracked to find another note, which included the relevant details.

Six days after the ransom note was found, a letter, mailed on August 19 from the city of Grantham, arrived in Lincolnshire. Most significantly, it began, "Game abandoned," and continued, "You will have to file your papers until I try again…as you know I never picked anyone up in Ipswitch [sic]." Nothing more was heard for two months, until the blackmailer decided on another game. In a letter to British Rail he demanded £200,000 ($330,000), and he threatened to derail a train. Once again, telephone contact failed.

On January 22, 1992, Bob Taylor was informed that Stephanie Slater, a real estate agent in Birmingham, had been

Michael Sams left a trail of envelopes for a ransom carrier to follow, after he alleged that he had kidnapped a prostitute in Ipswich. The first envelope to be discovered was attached to a white-painted brick, left under a bridge over a former railroad track.

Stephanie Slater, the real estate agent from Birmingham who was kidnapped by Sams on January 22, 1992. He released her, unharmed, a week later.

abducted while showing a potential client a house. The next day a letter to the real estate agency – matching the other letters in style – demanded £175,000 (nearly $300,000) to be delivered by the office manager, Kevin Watts. Because the kidnapping had occurred in Birmingham, the case had to be handled by West Midlands Police; the West Yorkshire Police were left to watch the investigation from the sidelines.

On the evening of January 29, Kevin Watts set off to follow a trail of messages that led him back and forth across the Pennines. His car was equipped with a radio to relay his position to the police, and he carried £175,000 in a canvas bag. Unfortunately, a thick fog developed, delaying him and frustrating police hopes of intercepting the kidnapper. Finally, at the end of a narrow lane, he left the money in a wooden tray on the parapet of a bridge that crossed the same disused rail track, just three miles from where the envelope and brick had been found. The police later realized that the kidnapper had waited below the bridge and then pulled the money down with a rope.

STEPHANIE IS SAFE

Four hours later, Stephanie staggered from a car near her parents' home, safe and unharmed. She had seen her kidnapper before he blindfolded her and was able not only to describe what he looked like, but also how he spoke. During her captivity, she had been driven in a car for many

hours and imprisoned for seven days in a tight-fitting coffinlike box.

The police had tape recordings of the offender's voice, which were broadcast on February 20 on the television program *Crimewatch UK* together with an artist's drawing from Stephanie's description. Within a short time a woman telephoned to say that the voice and drawing were those of her ex-husband, Michael Sams, who ran a tool repair workshop in Newark, in Nottinghamshire, some 30 miles from where Julie Dart's body had been dumped. And it was in the right-hand corner of Britton's triangle.

The next morning a police team was on its way to Newark. Sams greeted them at his workshop with the words: "I've been

Three years after his conviction for murder and kidnapping, Michael Sams – here without his glasses – was once more in court, charged with assaulting a prison officer.

THE RAILROAD CONNECTION

Both Bob Taylor and Paul Britton noticed the connection between bridges over railroad tracks and the places from which Sams made telephone calls and at which he arranged his pick-ups. He clearly had a detailed knowledge of the railroad system, and his threat to derail a train revealed expert technical knowledge.

Stephanie Slater described how he had worn a train badge on his duffel coat, and one of his wives reported that he was a train enthusiast. And at his home, there were big railroad signs all over the walls.

Bob Tayor reasoned that Sams's interest in railroads might provide a clue. He found a witness who had seen a man answering the offender's description some 600 yards from where Julie's body had been found on February 19, and Sams admitted having visited nearby Westby viaduct.

Better known as Stoke Summit, the viaduct is a famous place in British railroad history. It was here in 1938 that the steam locomotive *Mallard* reached a record-breaking speed of 126 mph.

expecting you." In the workshop and at his house a few miles away they found a wealth of incriminating evidence. He was arrested and confessed to the kidnapping of Stephanie Slater, but denied that he had murdered Julie Dart.

Born in Yorkshire, Michael Sams was 51 years old. He had completed his education at Hull Nautical College and served three years in the Merchant Navy before training in elevator installation and central heating. All three of his marriages had failed, as had his many business ventures. In 1978, he had been imprisoned after stealing and respraying a car. While in prison he developed cancer in his knee and his lower right leg was amputated. As in so many cases, Britton's profile had not led to Sams's identification, but he could nevertheless take pride in its accuracy.

Also, as Britton had predicted, Sams proved extremely devious under interrogation. He said "another man" had killed Julie and must have used his word

processor. Sams constantly changed his explanations for forensic evidence that was linked to him. At times he announced that he would answer no further questions, but then his desire to continue playing games would cause him to keep talking. And although the police recovered £19,000 (more than $31,000) from Sams's workshop and a further £5,000 ($8,000) that he had inadvertently dropped on the railroad track, he refused to admit that he had the rest, and where he had hidden it.

Taylor called on a Special Air Service (SAS) colonel with experience in detecting IRA arms caches to help him out. He also rented a radar device designed for probing ground. In the first week of December, more than nine months after the arrest of Sams, £120,000 (nearly $200,000) was dug up from two hiding places on the railroad embankment by Stoke Summit. Sams was found guilty of the murder of Julie Dart and subsequently admitted to all of his crimes in prison.

By now, Paul Britton was being consulted in a succession of rape and murder cases. These included the abduction and killing in Liverpool of two-and-a-half-year-old James Bulger by two young boys in 1993 and the case of Fred and Rosemary West. The Wests were arrested in 1994 for the murder of a succession of young girls, including one of their own daughters. However, a slightly earlier case was later to provoke bitter criticism of his methods. This was the investigation into the murder of Rachel Nickell.

CASE STUDY: MURDER ON WIMBLEDON COMMON

Early one morning in July 1992, 23-year-old Rachel Nickell took her two-year-old son and her dog for a walk on Wimbledon Common, in southwest London. Soon after, her murdered body was found beside a footpath. Her head had been nearly severed, 49 stab wounds had been inflicted, and her young son, covered in mud, was desperately trying to wake up his mother. The police had few leads. There was no forensic evidence such as blood,

The abduction and murder of two-year-old James Bulger by two boys in 1993 horrified the world. A security camera captured pictures of the toddler being led through a shopping mall. Paul Britton advised the police that the murder was not a game that had gotten out of hand, but rather that the abduction had been planned.

Rachel Nickell was brutally murdered in 1992 on Wimbledon Common, southwest London, while taking her dog and two-year-old son for a morning walk. Her body was discovered shortly after the murder, with her young son at her side, frantically trying to revive her.

semen, saliva, or hair. All they had to go on was a single unidentified footprint in the mud and the possibility that the murder weapon was a single-edged sheath knife with a brass hilt. The only lead came from passersby who had seen a young man washing his hands in a nearby stream.

The police in London invited Paul Britton to assist them. Britton asked for as much information as possible about the victim. He wanted to know if her appearance and personality made her a "high-level risk" or not. The information that he eventually obtained established that, although she was extremely attractive, she was not provocative and may have been unaware of the full effect of her visual impact on other people. Based on his considerable experience, Britton then began to draw up an image of the killer and his sexual fantasies.

"CHRISTIANS KEEP AWAY"

More than two months after the murder, a reconstruction of the crime was shown on the TV program *Crimewatch UK,* together with two pictures of men the police wished to interview. Within four hours, the police received more than 300 telephone calls. One name came up four times: Colin Stagg, a 29-year-old unemployed bachelor living alone less than a mile from Wimbledon Common.

Stagg had already been interviewed by the police but he was arrested at once on suspicion. The front door of his apartment was painted with a pair of ice-blue eyes and the inscription "Christians keep away, a pagan lives here," and inside detectives discovered various pornographic magazines and books on occult subjects. Investigators also discovered that a few

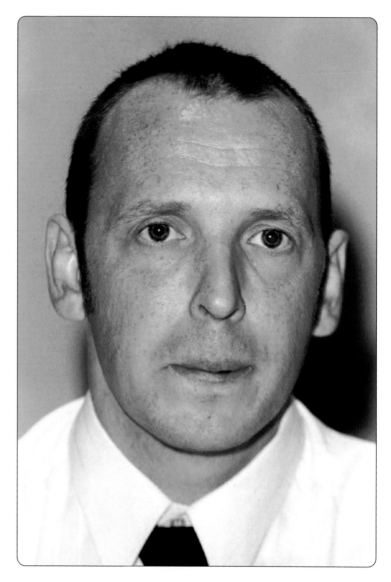

days after the murder he had been charged with indecent exposure on Wimbledon Common and fined.

In his interviews, Stagg consistently denied his guilt, although he admitted that he might have known the victim by sight and revealed a detailed knowledge of the common. Listening to tapes of the interviews, Britton concluded that although nothing clearly indicated Stagg's guilt,

A police file photograph of Colin Stagg, who became a suspect in the Rachel Nickell murder after being implicated by callers to the TV program **Crimewatch UK.**

On September 14, 1994, Colin Stagg left the court a free man, following the prosecution's unsuccessful attempt to convict him of the murder of Rachel Nickell.

BRITTON'S ANALYSIS

"After examination of the source material, I am of the opinion that the offender has a sexually deviant-based personality disturbance, detailed characteristics of which would be extremely uncommon in the general population, and would represent a very small subgroup within those men who suffer from general sexual deviation.

"I would also expect the offender's sexual fantasies to contain at least some of the following elements:

1: adult woman
2: the woman would be used as a sexual object for the gratification of the offender
3: there would be little evidence of intimate relationship building
4: there would be sadistic content; it would involve a knife or knives, physical control and verbal abuse
5: submission of female participant
6: it would involve anal and vaginal assault
7: it would involve the female participant exhibiting fear
8: I would expect the elements of sexual frenzy, which would culminate in the killing of the female participant."

nothing clearly indicated his innocence, either. Stagg was released after three days.

Not long afterward the police were given a letter that Stagg had written two years earlier in reply to a "lonely hearts" advertisement in a magazine: it detailed a masturbation fantasy. They again contacted Britton and asked if he could devise a covert operation that would further implicate "a person" in their inquiries or help to eliminate him.

Britton proposed that an undercover policewoman should contact the "person" by letter and in the course of the correspondence draw out his deepest sexual fantasies. Later, it might be necessary for her to meet with him. A member of Scotland Yard's covert operations unit, SO10, was selected. She was blonde and blue-eyed, aged 30, and was given the pseudonym "Lizzie James." The cover history that Britton drew up for her included early sexual abuse in an occult group, involvement in the ritual murder of a young woman, and the belief that she could achieve sexual satisfaction only with a man who had had similar experiences. The policewoman was to release this information only gradually, as the correspondence developed. Britton

In 2002, 10 years after Rachel Nickell's murder, Colin Stagg wrote a book, **Who Really Killed Rachel?**, *in which he claimed to identify the real killer.*

warned the police that this "sting" could take up to three months to establish any exchange of fantasies and a further two months to allow them to develop.

"LIZZIE" AND COLIN MEET

Lawyers of the Crown Prosecution Service gave their permission for the operation, although only after several months' consideration. However, Stagg's second letter to "Lizzie" after contact was made already included a sexual fantasy. His fourth letter particularly interested the

police because many of its details closely resembled the situation in which Rachel Nickell had been killed. More letters were exchanged until Stagg wrote: "I am going to make sure you are screaming in agony when I abuse you. I am going to destroy your self-esteem…." At the end of April 1993, the couple spoke on the telephone for the first time. Stagg referred obliquely to that fact that he continued to be a suspect in the murder but stated: "I want to tell you, like I told everybody…I never done anything…."

After a second telephone conversation, in which "Lizzie" – unwisely as it later turned out – said "quite frankly, Colin, it wouldn't matter to me if you had murdered her," and Stagg again protested his innocence, the couple agreed to meet for a picnic on his 30th birthday. A hidden tape recorder captured their conversation. It was inconclusive, but in a letter that Stagg handed to "Lizzie" as he left and in subsequent correspondence and meetings he revealed more fantasy details, which included the use of a knife. And at the end of July he told her that he had witnessed Rachel Nickell's murder. The details that he gave convinced the police that it was time to arrest and charge him.

Stagg's trial opened at the Central Criminal Court – the "Old Bailey" – on September 5, 1994. At the end of the following week, the judge ruled that all the letters and recorded conversations between "Lizzie" and Stagg were inadmissible as evidence. He said: "I am afraid this behavior betrays not merely an excessive zeal, but a substantial attempt to incriminate a suspect by positive and deceptive conduct of the grossest kind.... I would add that, even were I to be persuaded that this material could fairly and properly warrant the status of a confession, I regard it as so flimsy in its nature as to demand the conclusion that its potential prejudice clearly exceeds any probative value." The prosecution withdrew its case, Colin Stagg walked from the court a free man, and the murder of Rachel Nickell has remained unsolved.

> **Stagg's fourth letter to "Lizzie" included many details closely resembling the situation in which Rachel Nickell had been killed.**

The failure of the case against Colin Stagg and the judge's forceful comments on what was effectively entrapment of the suspect proved a major embarrassment for Paul Britton. And it resulted in Britton being called before a disciplinary hearing of the British Psychological Society in 2001.

POLICE APPROVAL

The early successes of David Canter and Paul Britton in their respective approaches led to an upsurge of police interest in the possibilities of offender profiling, and a subcommittee of the Association of Chief Police Officers (ACPO) was eventually established to look into the technique. Their conclusions were disturbing. Don Dovaston, then Assistant Chief Constable of Derbyshire, commented:

"There had been some less successful exploits than the Duffy case. Millions had been spent on the advice of people who thought they were doing the police a favor, when they weren't. We had psychologists saying the offender was a 6 foot 4 inch mentally disturbed black man who had recently arrived in the country, when the culprit turned out to be a 5 foot 8 inch white man who was perfectly stable and had lived here all his life. We began to delve into people's credentials, and found they often didn't have the level of expertise which was required, and the conclusion was that it was time to set up a list of accredited profilers."

A panel of some 20 approved psychology consultants was appointed, mostly psychologists and psychiatrists

The Police Staff College at Bramshill in Britain, the seat of the newly-established National Crime Faculty, which controls an approved panel of psychology consultants.

with experience of extreme criminal behavior. In 1995, it came under the control of the National Crime Faculty, a new central resource of information and expertise set up at the Police Staff College at Bramshill. Between 1990 and 2000, they were increasingly approached by police forces throughout the United Kingdom in cases of violent crime that were proving difficult to solve. However, as more than one investigator pointed out, they were but "one tool in the toolbox."

They frequently did little more than reassure the police about the direction that an inquiry was taking, rather than actively guide the police to identification. In most cases, forensic evidence or a successful surveillance led to the eventual arrest of the criminal.

A striking example was that of a man who sent letters to female flight attendants based at Gatwick airport in the south of England, demanding pornographic photos of them in their uniforms and threatening to disfigure them if they did not comply.

A consulting psychologist gave his opinion that the man was a low-level worker at the airport – possibly a cleaner or a van driver – and an unattractive, unmarried loner. When the offender was captured as the result of a surveillance operation, he turned out to be a 40-year-old senior engineer with British Airways, good-looking, previously married, and now living with a girlfriend. There was one positive aspect of the psychologist's intervention, however. The details the police gave reminded him of a similar incident he had consulted on earlier and in a different police jurisdiction. This link proved invaluable in moving forward the investigation, but it was an amazing piece of luck and highlighted the importance of establishing a national crime database.

INTRODUCING NATIONAL COMPUTERIZATION

The investigation of the case of the Yorkshire Ripper (see Chapter 3) lasted more than five years and resulted in the assembly of an overwhelming collection of information – handwritten or typed on file cards or sheets of paper, gathered into boxes, and virtually inaccessible. Following an inquiry in 1982 into the failure of existing systems for the collation of information, the British police set up their first centralized crime computer in

1987. It was originally called Home Office Major Enquiry System (HOMES), but somebody with a sense of humor realized that by adding the word "Large," its acronym would be highly appropriate. And so it was named HOLMES.

HOLMES revolutionized British detection. For the first time, the 43 separate police authorities in England and Wales, and those of Scotland and Northern Ireland, were able to establish communication at a national level. Systems were established that could check if serious crimes in different jurisdictions were part of a series. Training programs for senior investigating officers (SIOs) were instituted and specialist forensic support teams were set up.

As the computer system began dealing with serial murder, major disasters, and

Few profilers admit their mistakes publicly, but British psychologist Dr. Julian Boon confessed one significant failure. His profile of a man who wrote threatening letters to female flight attendants at Gatwick airport did not match, in any particular, the offender who was eventually identified.

THE MAGIC OF PROFILING

Many British police officers have publicly expressed skepticism of the value of offender profiling. As Tom Williamson, Deputy Chief Constable of Nottinghamshire, stated:

"There has been an approach of enterprising amateurism in the practice of profiling, and what we desperately need is to impose a discipline on that."

The failure of the prosecution in the Rachel Nickell case has highlighted the dangers of the police placing too much reliance on the advice and suggestions of a consultant psychologist with only a limited experience of legal practice and procedure.

However, considered only as "one tool in the toolbox," psychological analysis can be shown to be an important element in an inquiry. Much depends upon a good relationship being established between the consultant and the senior investigating officer, who must maintain at the same time an objective attitude, unswayed by a belief in the "magic" of profiling.

When police discovered the remains of three bodies in the Gloucester garden of Fred and Rosemary West, Paul Britton advised them: "He used the garden because the house is full."

serious fraud cases, a number of shortcomings in the original program became apparent. So, beginning in 1999, an updated system – HOLMES 2 – has been progressively introduced.

Derbyshire Police set up a separate data base. The most highly developed statistical approach to crime investigation in Britain so far, its acronym is very deliberate: CATCHEM (Centralized Analytical Team

Collating Homicide Expertise and Management). This complex system contains some 9,000 details of homicides committed since 1960. The program has proved invaluable: in one case, which analyzed the age, employment, forensic history, and place of residence of a host of possible suspects, 150 individuals were highlighted, and the 24th of these was positively identified by DNA.

A trial system recently introduced in certain areas by British police uses surveillance cameras, linked to computer files, to identify and track known criminals observed on the streets.

BEHAVIORAL EVIDENCE ANALYSIS

> **"I learned an important lesson...That offenders lie,"** Brent Turvey wrote. **"The only way to get an objective record of the behavior that occurs in a crime scene between the victim and the offender is through the documentation and subsequent reconstruction of forensic evidence."**

Following the claimed successes of the FBI's BSU, a steadily growing number of freelance profilers – both academic psychologists and independent practitioners – have appeared in the United States. One of the most active is Brent Turvey, a forensic psychologist in private practice in California. He first became interested in the psychology of sex offenders as a teenager, when he learned of the incest and molestation that his then girlfriend had suffered. He decided to make the subject his study. In 1991, while he was still an undergraduate in psychology at Portland State University, he had his first interview with the serial murderer Jerome Brudos.

CASE STUDY: JEROME BRUDOS

Jerry Brudos, born in 1939, grew up hating his domineering mother. Yet from the age of five he began to develop a fetish for women's clothing, particularly their shoes. He took his first pair – which his mother then burned – from the town dump. After this he began to steal from his sister and later from neighbors, sometimes also taking underwear from their clotheslines.

At 17 he forced a girl at knifepoint to strip while he took photographs. As a result, he was defined as suffering from "early personality disorder," and ordered to attend a local mental hospital for therapy

When Karen Sprinker was abducted, two people reported earlier sightings of a large woman loitering nearby. This might have been Brudos, dressed in women's clothing, but following his confession his wife, Ralphene (pictured), was arrested too.

as an outpatient. In March 1959, he joined the U.S. Army, but was discharged after six months when he told an army psychiatrist that a beautiful Korean girl crept into his bed every night. Back home in Oregon, living in his family's toolshed, he began to knock women down in the street and steal their shoes.

These attacks ceased in 1961 when Brudos met a young woman, got her pregnant, and married her. For a time he

seemed to have settled down, but in 1967 – while his wife was in the hospital with their second child – he started to steal again. Brudos followed a girl wearing attractive shoes, broke into her apartment, choked her unconscious, raped her, and carried off her shoes. It was at this time that Brudos began to make his wife walk around their house naked, while he took photographs; he also posed for himself, wearing women's underwear.

MURDER AND MUTILATION

Then, on January 26, 1968, 19-year-old Linda Slawson came to his door, selling encyclopedias. He took her into his garage, beat her, raped her, and then strangled her to death. While he sent his wife out for hamburgers, he dressed up Linda's body in the clothes he had been collecting; he then cut off her left foot, locked it in the freezer, and dumped her body in the nearby Willamette River.

In July, 16-year-old Stephanie Vikko disappeared, and her remains were not discovered until the following March. On November 26, 23-year-old Jan Whitney disappeared. Brudos had picked her up after her car broke down and taken her to his garage where he killed her and hung her from a hook, leaving the body, dressed to his taste, for several days. He disposed of her corpse in the river, but not before cutting off one of her breasts as a trophy.

Four months later, on March 27, 1969, Brudos abducted 19-year-old Karen Sprinker from a department store parking lot. At home he raped her, forced her to pose for photographs, then killed her by hanging her. He then sliced off both her breasts and dumped the remains of her body in the Long Tom River. His final victim was 22-year-old Linda Salee. On April 23, he pretended to be a police officer and "arrested" her on suspicion of shoplifting before killing her. "Her breasts were all pink – the nipples weren't dark like they should be.... I didn't cut them off because they didn't appeal to me," he said later. He tried to make plaster casts of them instead.

All this time the police had been making inquiries into unexplained disappearances, but had no proof of homicide. On May 10, fishermen found Linda Salee's body, weighed down with a car transmission, in the Long Tom River. Two days later, Karen Sprinker's body was discovered in the same river by a team of divers. Interviews with local co-eds brought out stories of a strange "Vietnam vet" who hung around the campus. Jerry Brudos was identified on May 25, questioned, and charged five days later.

A search of the Brudos home uncovered a large collection of women's shoes and clothing, together with hundreds of photographs. One constituted damning evidence: Brudos had photographed one of his victims, carefully dressed by him, hanging from the hook in his garage. He had placed a mirror beneath her skirt – and in doing so had inadvertently caught a reflection of himself.

At his trial Brudos at first pleaded insanity, but changed his plea after seven doctors declared him sane – though

Jerome Brudos, like more than a few serial killers, pretended to be a plainclothes cop. He stole women's clothing and dressed his dead victims to his taste before photographing them. He also cut off their breasts and kept them as trophies.

suffering from "severe personality disorders." He was found guilty of four murders and sentenced to life imprisonment.

INTERVIEW WITH BRUDOS

Turvey talks about his interview with Brudos in his book *Criminal Profiling* (1999):

"I realized how truly naive my understanding of sex offenders was. I spent five hours with him, and he lied to me almost the entire time. He lied about almost everything he had ever done (or rather, he claimed, everything that he hadn't done). The only reason that I was not completely taken in by his charming personality and generous, affable nature was the fact that I had reviewed the investigative file. Prior to the interview, I had gone to the Marion County Sheriff's office in Salem, Oregon. I had read autopsy reports on all of the victims,

> **A search of the Brudos home uncovered a large collection of women's shoes and clothing, together with hundreds of photographs.**

Brent E. Turvey has developed his own system of criminal profiling distinct from that of the FBI, which he has named "behavioral evidence analysis."

Deciding to supplement his psychology degree, Turvey applied to and was accepted in the Graduate Forensic Science Program at the University of New Haven, in Connecticut. Subsequently, he has developed a personal approach to criminal profiling that he has named "behavioral evidence analysis."

INDUCTIVE REASONING

Turvey is opposed to the assembly and analysis of statistical data in criminal profiling that is currently the practice of the FBI and (in a different way) of David Canter and his colleagues in the United Kingdom. In principle, he believes in the independent assessment of each individual case – not intuitively, but after scientific consideration of all the available evidence. (Nevertheless, despite his emphasis on scientific rigor, it is arguable that his analysis will inevitably be influenced by his experience of previous cases. As he himself states, the final stage of his profiling method is "a matter of expertise and not a science.")

He characterizes the FBI approach as essentially inductive: that is, it begins from a number of known premises and reaches a (more or less) logical conclusion. There is nothing fundamentally wrong with inductive reasoning, but it can be used to reach a positive conclusion that may be misleading. Turvey quotes a typical example:

looked at the crime scene photos, and read the investigators' reports. I had even seen some of the photos Jerry had taken of himself posing with his victims.

"I learned an important lesson through that experience. The lesson was that offenders lie. The only way to get an objective record of the behavior that occurs in a crime scene between a victim and the offender is through the documentation and subsequent reconstruction of forensic evidence. It's a lesson that I took to heart."

PREMISE: Most known serial murderers are Caucasian.
PREMISE: Most known serial murderers are male.
PREMISE: Most known serial murderers operate within a "comfort zone."

"YOU KNOW MY METHODS, WATSON"

Although Sherlock Holmes is generally credited with achieving his startling conclusions by deduction, his methods were in fact inductive.

Dr. Watson: "Now, I have here a watch which has recently come into my possession. Would you have the kindness to let me have an opinion upon the character of the late owner?"

Holmes: "He was a man of untidy habits – very untidy and careless. He was left with good prospects, but he threw away his chances, lived for some time in poverty with occasional short intervals of prosperity, and finally, taking to drink, he died."

"How in the name of all that is wonderful did you get these facts? They are absolutely correct in every particular."

"Ah, that is good luck. I could only say what was the balance of probability. I did not at all expect to be accurate."

"But it was not mere guesswork?"

"No, no: I never guess. It is a shocking habit – destructive to the logical faculty. What seems strange to you is only so because you do not follow my train of thought or observe the small facts upon which large inferences depend.

"When you observe the lower part of that watch case you notice that it is not only dinted in two places but it is cut and marked all over from the habit of keeping other hard objects, such as coins or keys, in the same pocket. Surely it is no great feat to assume that a man who treats a fifty-guinea watch so cavalierly must be a careless man. Neither is it a very far-fetched inference that a man who inherits one article of such value is pretty well provided for in other respects.

"It is very customary for pawnbrokers in England, when they take a watch, to scratch the numbers of the ticket with a pin-point upon the inside of the case.... There are no less than four such numbers visible. Inference – that [he] was often at low water. Secondary inference – that he had occasional bursts of prosperity.... Finally, I ask you to look at the inner plate, which contains the keyhole. Look at the thousands of scratches all round the hole – marks where the key has slipped... you will never see a drunkard's watch without them...."

(Sir Arthur Conan Doyle, *The Sign of Four*, 1888)

The popular image of the fictional Sherlock Holmes, who frequently remarked "this is a three-pipe problem," was established very early in the magazine illustrations which accompanied Conan Doyle's stories.

TURVEY'S DEDUCTIVE METHOD

Among the basic assumptions of the deductive method, Turvey lists:
- **No offender acts without motivation.**
- **Every single offense should be investigated as having unique behavioral and motivational characteristics.**
- **Different offenders can exhibit similar behavior for completely different reasons.**

- **No two cases can be completely alike.**
- **Human behavior develops uniquely in response to environmental and biological factors.**
- **Criminal MO can evolve over time and over the commission of multiple offenses.**
- **A single offender is capable of multiple motives over the commission of multiple offenses or even during the commission of a single offense.**

CONCLUSION: It is *likely* that a serial murderer will be a male Caucasian operating within a comfort zone.

The critical word here is "likely." Turvey then gives an example of the sort of conclusion that is reached by a typical profiler: the given serial murderer *will be* a male Caucasian operating within a comfort zone. The other critical word in this argument is "known." Turvey asserts that it is unscientific to reason in this way from the general to the specific – and equally from the specific to the general.

As an example of the inherent dangers of inductive generalization, Turvey puts forward the following scenario:

"A 24-year-old white female is raped in her apartment on the first floor. A detective working on the case goes to the press and states that there is a sexual predator on the loose who only attacks white females who live in first-floor apartments."

As he points out: "Not only is it possible for the same rapist to attack victims in other locations based on opportunity, but also an offender could read this statement in the news and make it a point not to attack victims in first-floor apartments ever again."

There are additional drawbacks to inductive profiling. First, the statistical data is drawn from limited population samples and may not be applicable to a specific offender. Second, the data is derived from apprehended offenders. As Turvey points out, it "cannot fully or accurately take into account current

The first stage in Brent Turvey's behavioral evidence analysis, named "equivocal forensic analysis," is the study of photographs, videos, and investigators' reports in relation to the crime scene and autopsy. Here, FBI crime scene investigators gather vital evidence.

Brent Turvey claims that he places greater emphasis on the victim than the FBI profilers. An important element is the assessment of the degree of risk taken by the victim at the moment of abduction. As experience shows, prostitutes are among those at greatest risk.

offenders who are at large, therefore it is by its very nature missing data sets from the most successful or skillful criminal populations." Third, an inductive criminal profile can often contain inaccuracies – which have on a number of documented occasions implicated innocent people.

TURVEY'S DEDUCTIVE METHOD

Deductive reasoning, Turvey points out, involves arguments where, if the premises are true, the conclusions must also be true. His "deductive criminal profile" involves reasoning "in which conclusions about

offender characteristics follow directly from the premises presented and where the profile itself is concerned with a specific pattern of behavior as opposed to reasoning from the characteristics of an average offender type."

It is clear that Turvey starts from a very different basis from that of the FBI's behavioral profilers, who base their analysis on previous cases. As he writes: "Statistical generalizations and experiential theorizing, while sometimes initially helpful, are incomplete and can ultimately mislead an investigation."

THE FOUR STAGES

Behavioral evidence analysis – which is deductive in style – is divided into four stages. The first is named "equivocal forensic analysis." It is "equivocal" in the sense that the evidence can have more than one interpretation and the analysis is directed to assessing the most likely interpretation. The forensic evidence should include (but is not limited to): crime scene photos, videos and sketches; investigators' reports; evidence logs and submission forms; autopsy reports, videos, and photos; interviews with witnesses and neighbors; a map of the victim's movements prior to the crime; and the background of the victim.

Turvey's method places great emphasis on the characteristics of the victim – frequently insufficiently considered in criminal profiling – and the second stage of the analysis is called "victimology." Knowing how, where, when, and why a particular victim was chosen can tell a great deal about the offender. For example, the physical build of the victim can be indicative of the build of the offender. In a

> Knowing how, where, when, and why a particular victim was chosen can tell the experienced profiler a lot about the offender.

similar way, if the victim was abducted without a struggle and is described as naturally wary, it suggests that the offender was known to the victim or was socially adept in persuading her to accompany him.

Part of victimology is risk assessment. The profiler is interested not only in the amount of risk that the victim's lifestyle places her in, but also the risk she was in at the time of the attack and the risk that the offender was prepared to take.

The third stage is known as "crime scene characteristics." These are "the distinguishing features of a crime scene as evidenced by an offender's behavioral decisions regarding the victim and the offense location, and their subsequent meaning to the offender." These distinguishing features include method of approach, method of attack, method of control, location type, nature and sequence of sexual acts, materials used, verbal activity, and precautionary acts. They help the profiler distinguish between MO and signature, as well as allowing inferences to be made as to the offender's state of mind, planning, fantasy, and motivation.

IDENTIFYING MOTIVATION

In his book *Criminal Profiling*, Brent Turvey provides an example of how tricky it is to determine the motives for an offender's behavior. He outlines a scenario in which a rapist attacks a victim in a public park: the rapist pulls the victim's shirt up over her face and leaves it there for the duration of the attack.

What was the intended result of this action? Was it to cover the victim's eyes, perhaps to prevent identification of the offender? To cover the victim's face, so as to render her anonymous? To trap the victim's arms? To expose the victim's breasts? Or is this the expression of some kind of displaced fantasy?

Much of this – with the exception of the victimology – is material that will equally be taken into account by profilers with access to profiling systems such as VICAP computer analysis.

Turvey stresses that the fourth stage of his approach – "offender characteristics" – should not be considered a final conclusion, but should constantly be updated and reviewed as new evidence comes to light or old material is discredited.

Among the offender characteristics that Turvey and his colleagues claim to be able to distinguish are: physical build; sex; work status and habits; feelings of remorse or guilt; vehicle type; criminal history; skill level; aggressiveness; residence in relation to the crime; medical history; marital status; and race. These are conclusions that many other profilers – whether working from statistical data or more intuitively – also feel justified in reaching.

Turvey's behavioral evidence analysis comes to similar conclusions as those of other profiling methods, but by a different route. By insisting on the uniqueness of each case, he avoids the pitfalls of applying statistical averages to a specific case. In fact, numerous commentators have remarked upon the "sameness" of many of the FBI's profiles.

The strength of Turvey's profiles lies in the painstaking detail of the assessment of each individual crime, which is reflected in the reports provided to the investigators.

Brent Turvey cites a number of cases in which conventional profiling can do more harm than good. One was that of Richard Jewell (above), a security guard who reported the discovery of a bomb at the 1996 Olympic Games in Atlanta. The national media reported that FBI profiling had identified Jewell as the prime suspect, but he was never charged, and the FBI eventually cleared him of all suspicion.

GEOGRAPHIC PROFILING

Mapping the points at which violent crimes take place has proved of inestimable value in tracking down the offender. A study of the activities of the "Rostov Ripper," Andrei Chikatilo (left), narrowed the search to the local railroad network. When he was apprehended, he confessed to 55 killings.

Most investigators agree that the experienced offender reveals a psychological makeup that is distinguishably different from what we consider "normal" and can provide clues to his eventual identification, as previous chapters have shown.

In addition to a characteristic MO, the criminal, like a predatory animal, also has his "patch" – the area within which he operates. At first glance this may appear dauntingly large, but investigators can usually detect a pattern within this zone – as David Canter's first case, the analysis of John Francis Duffy's rapes and murders, clearly demonstrated.

The first attempt to investigate geographical patterns in the incidence of crime was made by two French criminologists, André-Michel Guerry and Lambert-Adolphe Quetelet, in the mid-19th century. They drew up maps of the occurrence of violent and property crimes throughout France, examining how they related to levels of poverty in different areas.

Early in the 20th century, sociologists from the University of Chicago carried out a similar study in the windy city. However, it was not until recently – with the wide availability of complex computer programs – that a technically more sophisticated development came about in the mapping and detection of crime.

PINPOINTING THE YORKSHIRE RIPPER

The specific application of mapping to the analysis of violent serial crime has come into particular prominence in the past 20 years. An intuitive awareness of the offender's geographic pattern of behavior, based solely on many years of practical experience, however, has played a significant part in the investigator's approach for a long time.

In *The Scientific Investigation of Crime* (1987), Stuart Kind – the former director of the Home Office Central Research Establishment at Wetherby, Yorkshire – describes a high-level meeting that took place on December 1, 1980, to provide advice in the investigation into the case of the Yorkshire Ripper. Among the murders looked at by the advisory panel was that of a 20-year-old student at Bradford University, who had been killed on September 1, 1979. One member of the panel, Commander Ronald Harvey from New Scotland Yard, studied the evidence and then exclaimed: "Chummy lives in Bradford, and he did it going home!"

Kind explains, in a graphic example (strikingly similar to the description in Chapter 5 of the plotting of behavioral "maps"), how this intuitive revelation was further examined by the panel.

CONNECTING THE THREADS

"Consider a map upon which we plot the positions of the 17 assumed Yorkshire Ripper crimes. If we mark each of these positions with a map pin and tie a piece of thread to it we are then in a position to consider the following question: 'At which single location on the map could be placed an 18th pin, such that if we stretched the 17 threads and tied each thread to the 18th pin, the minimum total amount of thread would be used?'"

The actual location of what Kind calls this "center of gravity" was carried out not with pins and threads but by computer, first for all 17 locations and subsequently for a smaller number of crimes that might also be attributable to the Yorkshire Ripper. In all cases the computer pinpointed a location close to the city of Bradford, "possibly in the Manningham or Shipley area." Kind points out that this analysis did not take account of the actual distances by road, but only "as the crow flies." Nevertheless, when Peter Sutcliffe was charged with the murders only a month later on January 2, 1981, he was found to be a native of Bradford and resident in a district between Manningham and Shipley.

The computer programs that have now been developed use software known as GIS (geographic information system), which can correlate both space and time factors. The International Association of Crime Analysts (IACA) has calculated that the demand for GIS experts in law enforcement has grown tenfold over the past 15 years.

As Dr. Nancy La Vigne, director of the U.S. National Institute of Justice's Crime Mapping Research Center, said: "It's human nature to respond to this kind of graphical representation. What you get is a

> **Commander Ronald Harvey studied the evidence and then exclaimed: "Chummy lives in Bradford, and he did it going home!"**

far more sophisticated understanding of what's happening on the streets."

One valuable aspect of GIS mapping is the ability to identify crime "hot spots" rapidly. Various police forces developed their own ways of doing this. A very early system developed in Illinois was known as STAC (Spatial and Temporal Analysis of Crime); the New York Police Department made successful use of their "CompStat" process; and two counties of New York State operate "GeoMIND" (Geographically-linked Multi-Agency

Information Network and Deconfliction) to assist police in making decisions.

About ten years ago police officers suggested that investigators could assess the solvability of current serial murder cases by using GIS to analyze previous known cases. In the case of the "Hillside Stranglers," for example, retrospective analysis provided confirmation of the original police approach.

As the analysis of the geographical distribution of serial murder has developed, researchers have attempted to

The Yorkshire Ripper's den turned out to be a disorderly mess of empty bottles – but it proved to be at a location central to all of the principal murders.

THE "HILLSIDE STRANGLERS"

From October to December 1977, nine part-time prostitutes were murdered – strangled to death – within a small area of Los Angeles, and another was killed in a similar manner in February 1978. The young women had all been tied and raped. Their naked bodies had been cleaned after death, leaving few clues, and they were usually left openly on hillsides close to local police stations.

At first the police assumed that a single killer was responsible, but forensic evidence eventually established that two men were involved. Correctly assuming – as it later transpired – that the victims had been murdered on the premises of one of the offenders, Los Angeles Police Department computer analysts drew up a map based on the points where the women had been abducted, where their bodies were left, and the distances between them. The computer identified an area of just over 3 square miles, so the LAPD saturated this "sphere of concern" with 200 officers in the hope of apprehending the killers in the course of their abduction of another victim.

Whether it was this concentrated police presence or disgust at the living conditions of his fellow murderer is still open to question, but Kenneth Bianchi, one of the killers subsequently identified, moved to Bellingham, Washington. Arrested there for the rape and strangling of two college students, he implicated his cousin Angelo Buono in the earlier crimes. The LAPD later realized that the center of their map lay very close to Buono's automobile upholstery shop in Glendale, a suburb of Los Angeles.

Angelo Buono, implicated with his cousin Kenneth Bianchi in the "Hillside Strangler" murders of Los Angeles prostitutes. All but one occurred in a circle close to the home the two men shared.

Two criminologists later carried out a retrospective analysis of the case. They found that the center of the sites where bodies were dumped lay close to Buono's home – but there was an area immediately surrounding this center where no crimes occurred, suggesting that Buono did not want to kill too close to his home.

make a number of generalizations based on the information available from solved cases. For example, in 1990, one university researcher estimated that, if the body is dumped at a site different from that of the murder, the killer generally lives in the area where the original attack occurred.

On the other hand, if the body is left at the murder scene, it is possible that the killer is not local.

A crime scene close to a major road can also indicate that the murderer is not very familiar with the area; however, if the crime scene is a mile or more from a major road it suggests that the killer is local. A hidden body can mean that the offender intends to use the dump site again, a further indication that he resides locally. However, a body left unconcealed suggests that the killer is unconcerned whether or not it is discovered, suggesting that he is a transient. Generalizations such as this, however, can mislead investigators, and there are countless exceptions. "Disorganized" killers, for example, tend to live close to where they commit their crimes and usually leave the victim's body at the murder scene. "Organized" killers may roam far and wide and leave the bodies of their victims in isolated spots where they often remain undiscovered for months.

Kenneth Bianchi, one of the "Hillside Stranglers," in court during his trial. He was finally arrested after he had relocated to Bellingham, Washington, where he raped and strangled two college students.

The Rostov Ripper, Andrei Chikatilo, and the railroad map that led to his apprehension. He had been questioned several times as a suspect but was released each time, following the protests of local Communist Party officials.

THE ROSTOV RIPPER

Starting in November 1978 and continuing for the next 12 years, dozens of corpses – young women and children of both sexes – were discovered in forests near train and bus depots in the Rostov-on-Don region of Russia. The victims had been brutally raped and killed, then stabbed repeatedly, especially around the face. Some had had their tongues bitten off; others were disemboweled.

In August 1984 alone, eight victims were discovered. A man named Andrei Chikatilo was questioned several times as a suspect, but local Communist Party officials protested because he was known as a faithful and active party member.

Investigators came to the conclusion that the offender used local commuter trains to target his victims and then lured them into the nearby forest. The local railroad network offered a map of the killer's area of activity but – unlike most maps of similar crimes – it was long and narrow without an obvious center. Investigators therefore decided to drive the offender into a trap in "Operation Forest Strip."

More than 300 uniformed police were put on duty in a very obvious way in all the train stations along the line – except for three, which were covered by plainclothes officers. In November 1990, Chikatilo was spotted in one of the stations with blood on his face and hands. Another corpse was found nearby, and he was arrested. He confessed to 55 killings and gave the police details of the mutilations he had performed. In October 1992, Chikatilo was convicted of 53 counts of murder and sentenced to death. President Boris Yeltsin rejected an appeal for clemency and Chikatilo was executed by gunshot on February 15, 1994.

THE THRILL OF THE HUNT

Dr. Kim Rossmo, a detective inspector in Vancouver, British Columbia, has developed the technique of geographic profiling in violent crime to its most advanced stage. It is a scientifically structured statistical approach and has little direct concern with the psychology of the offender. As an epigraph to his book on the subject (*Geographic Profiling*, 2000), Rossmo quotes former FBI agent John Douglas: "Interview the subjects: what they'll tell you is, the thing that was really appealing to them was the hunt, the hunt and trying to look for the vulnerable victim."

Westley Allan Dodd, for example, who was found guilty of the murder of three children in Washington in 1989, wrote in his diary: "Now ready for my second day of the hunt. Will start at about 10:00 A.M. and take a lunch so I don't have to return home." And, at the same time, he expressed concern that should he kill a child in the park where he planned to hunt, he would lose his "hunting ground for up to two or three months."

Geographic profiling, therefore, writes Rossmo, is devoted to an analysis of "the spatial patterns produced by the hunting behavior and target locations of serial violent criminals.... Serial murder, for example, includes victim encounter, attack, murder, and body dump sites. The patterns and methods of offender hunting activity are analyzed from a geography of crime perspective. By establishing these patterns,

Westley Allan Dodd kept a diary in which he described how he set out to hunt for his victims.

it is possible to outline, through analyzing the locations of the crimes, the most probable area of offender residence." Rossmo is careful to stress, however, that the technique "does not solve crimes – that is the responsibility of the assigned investigator." It should also be said that this concept of geographical profiling runs counter to the testimony that Ann Rule gave before a U.S. Senate subcommittee in 1983, that serial killers – at least in the United States – frequently travel long distances to hunt for victims.

SERIAL KILLER CATEGORIES

Rossmo states that serial killers can be divided into four types according to the way in which they find their victims:

Hunter: Sets out specifically to search for a victim, basing the search from his residence.

Poacher: Sets out specifically to search,

Kim Rossmo's system of geographic profiling is designed to analyze the spatial patterns of the offender's hunting behavior.

but from an activity site other than his residence, or commutes or travels to another location during the search.

Troller: Opportunistically encounters a victim while involved in other non-predatory activities.

Trapper: Takes up a position or occupation or creates a situation that allows him to encounter victims within a location under his or her control.

Rossmo also defines three types of attacker:

Raptor: Attacks a victim almost immediately on encounter.

Stalker: First follows a victim upon encounter, gradually moving closer, waiting for an opportunity to attack.

Ambusher: Attacks the victim after he or she has been enticed to a location, such as a residence or workplace, controlled by the offender. The victim's body is frequently concealed in the same location.

According to Rossmo: "This typology is remarkably similar to Schaller's (1972) description of certain hunting methods used by lions in the Serengeti...."

"Hunter" and "poacher" are similar to the descriptions "marauder" and "commuter" used by David Canter in a study of serial rape in England. Marauders are individuals whose homes lie close to the center of the rough circle that encloses their crimes while commuters travel into another area. Out of 45 serial rapists considered in Canter's study, only five were classified as commuters.

On the other hand, a 1993 study by the FBI found that half of 76 serial rapists lived outside the circle of their crimes. They concluded that the disparity could be attributed to differences in European and American urban structure. Geographical

MENTAL MAPS

All humans develop mental maps, images of familiar areas such as neighborhoods or cities that are stored in the memory. As well as spatial information, these images include such details as color, sound, feeling, sentiment, and significant symbols. The spatial elements are divided into five types:

1: PATHS: Routes of travel that tend to dominate most people's images of cities and other central locations, such as highways or railroads.

2: EDGES: Boundaries, such as rivers, railroad tracks, or major highways.

3: DISTRICTS: Sub-areas with recognizable defining characteristics and with well-established centers but vague borders, such as financial districts, ethnic quarters, or "skid rows."

4: NODES: Centers of intense activity, such as major road intersections, railroad stations, or corner stores.

5: LANDMARKS: Recognizable symbols that are used for orientation, such as road signs, billboards, trees, or tall buildings.

considerations can also affect the formation of a crime circle. For example, the "Werewolf Rapist," José Rodrigues, found guilty of a long string of sexual assaults in 1990, lived in Bexhill on the south coast of England. As he obviously could not pursue his criminal activities southward into the English Channel, his crimes lay only within a rough half-circle to the north of Bexhill.

Geographical profiling is based on the premise that most people have an "anchor point." For the majority this is their home, but it may also be a work site or the home of a close friend.

> "Where we go depends upon what we know.... What we know depends upon where we go," wrote David Canter in *Criminal Shadows*.

Some criminals may base their activities on a social center, such as a bar or pool hall. This anchor point is certain to be within the person's mental map, and is likely to be close to the center. As criminologists have pointed out: "Very few criminals appear to blaze trails into new, unknown territories or situations, in search of criminal opportunities."

The question of the identity of Jack the Ripper continues to attract the attention of criminologists, including David Canter, who applied the concept of the anchor point to the problem. He took up the suggestion offered by historian Paul Begg that the prime suspect was Aaron Kosminksi – a theory with which the FBI profilers agreed (see Chapter 3). Although Begg did not know Kosminski's address he was able to establish that of his brother Wolf, who had taken responsibility for him after his committal to a mental asylum. Begg thought it likely that Aaron lived near by – in what the Assistant Chief Constable at the time had described as "the very heart of the district where the murders were committed."

Arguing that "to maintain the optimum distance that balances familiarity and risk, you would have to commit your crimes in a circular region around your home," Canter drew up a map of the Whitechapel area, dotted with the sites of the Ripper murders. Wolf Kosminski's residence lay roughly at their center.

GEOGRAPHIC TARGETING

Working with Simon Fraser University in Vancouver, Kim Rossmo and his colleagues have developed a "criminal geographic targeting" (CGT) computer system. Using statistical formulas that define the relative probabilities of the distances of the crimes from the offender's primary anchor point, the computer program produces a three-dimensional representation in color that is called a "jeopardy surface" (defined as areas where crimes are likely to occur).

This can then be projected onto a street map – in the same way that an ordinary relief map indicates changes in altitude by colored contours – to produce a complex and colorful "geoprofile." The peak area – the "high ground" in the map – will be an indication of the locality of the offender's anchor point.

The area of East London in which Jack the Ripper carried out his series of murders, and the site of the home of Wolf Kosminski. The street lamps mark the position of the five murders definitely attributed to the Ripper; Kosminski's home was central to all of them.

An artist's impression of the Mardi Gra (sic) bombers in court – Edgar Pearce (left) and his older brother Ronald Pearce.

The geoprofile can sometimes produce more than one peak area, an indication that the offender has more than one anchor point. Between 1994 and 1998 in Britain, the "Mardi Gra Bomber" was responsible for placing 36 explosive devices mostly in the Greater London area, which were accompanied by ransom demands. These were mailed or otherwise delivered to a wide variety of targets: automated teller machines, supermarkets, pay phones, business premises, and private residences. Scotland Yard requested a geoprofile from the Canadian profilers, which produced two peak areas: a primary peak around Chiswick in west London and a secondary peak in southeast London.

When 61-year-old Edgar Pearce and his 67-year-old brother Ronald were arrested while trying to withdraw ransom money from an automated teller machine, police discovered that while they both lived in the Chiswick area they also had family in southeast London.

Although geographic profiling was first developed for the analysis of serial crime, Rossmo cites a Canadian case in which the anchor point of a man who committed a single murder – admittedly followed by a series of bizarre actions – was pinpointed by CGT. One night in October 1995, a man

A police file photograph showing the face of Edgar Pearce. He was sentenced to 224 years at the Old Bailey on April 14 , 1999, after a three-year campaign of terror and blackmail against Barclay's Bank and the Sainsbury's supermarket chain.

RIGEL

A specific geographic-profiling computer workstation has been developed in Vancouver by Environmental Criminology Research Incorporated. It is named "Rigel" after the blue super giant star that forms the "heel" of the constellation of Orion the Hunter.

The developers say that it is "designed to support the hunter – the police detective – in his or her efforts to apprehend criminal offenders, just as Rigel the star supports Orion."

The computer carries out a million or more calculations in a typical analysis. Crime locations, broken down by type, are entered as either specific addresses or in latitude and longitude obtained as GPS (global positioning system) readings. Following this, various "scenarios" are created and examined. Geoprofiles and jeopardy surfaces (that is, areas where crimes are most likely to occur) can be rotated and viewed from different aspects to assist interpretation, and digital photographs of the peak area can be overlaid.

Large databases, such as sex offender registers, case management programs, or ViCLAS, can then be accessed so that the police can make best use of their relatively limited resources.

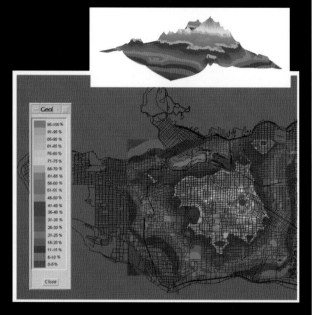

The Rigel software, which has a 70 percent accuracy rate, presents information in the form of a 2-D or 3-D result value map showing the most probable locations of the center of criminal activity. In this 3-D image of downtown Vancouver, the dark red area indicates the street where the offender most likely lives.

wielding a baseball bat attacked two teenage girls in British Columbia. One was killed and her body dumped 20 miles away; the other managed to stagger to a nearby hospital. A few days later, the killer made a number of taunting telephone calls to the dead victim's family and subsequently stole and defaced her gravestone. Finally, he wrapped a note around a wrench and threw it through a house window. With a total of 13 such incidents, CGT was able to indicate the anchor point within a small area, and the man was soon apprehended.

Geographic profiling has also resulted in the release of an innocent man. Early one

morning in January 1969, a young nursing assistant in Saskatoon, Saskatchewan, tried to catch a bus to work. But she was dragged into an alley, raped, and stabbed to death. Sixteen-year-old David Milgaard was convicted of her murder and sentenced to life in prison. He maintained his innocence throughout many years of incarceration.

In 1990 an alternative killer was suspected. He was Larry Fisher, a serial rapist who lived only a block away from the bus stop. Milgaard's family obtained a review of his case and a geographic assessment was undertaken.

"There were strong parallels, in MO and

crime site, between Fisher's rapes and the murder," Rossmo has written. "Same immediate area, identical location type (alleys protected from observation by garages, fencing, and vegetation), similar hunting style, same attack method, clothing manipulation, use of a knife, and brutality of sexual assault."

In addition, Rossmo says, "The circumstances of Milgaard on the morning of the murder did not support an opportunity to commit the crime." Finally, in 1997, DNA analysis resulted in Milgaard's exoneration and the arrest of Fisher for the murder.

The use of geographic profiling is spreading rapidly. In the past few years, the Vancouver Police Department has received requests for its services from the United States, Britain, Germany, Belgium, Greece, South Africa, Mexico, Australia, New Zealand, and several countries in the Middle East. To date it has assisted in more than one hundred investigations involving over 1,500 incidents for the FBI, Scotland Yard, and others. The service is now available in Britain through the National Crime Faculty, a police organization that provides advice and expertise in criminal investigations.

The Stony Mountain penitentiary in Winnipeg, Manitoba, where David Milgaard was confined for more than 20 years, until a geographic profiling assessment and a DNA analysis established his innocence.

THE CRIMINAL WORD

Criminals frequently feel compelled to communicate with their pursuers, by letter or telephone. Their messages may be taunting and confident, or a desperate appeal for help – like that left by teenage killer William Heirens (left) at the scene of one murder.

Paul Britton has rightly described himself as "the jigsaw man" because he puts together scraps of behavioral evidence and builds them into a plausible picture of the offender's psychological makeup. However, in some cases – particularly demands for ransom in kidnappings or "poison pen" letters – the investigator and the profiler can come closer to the physical presence of the criminal. They possess written or even spoken communications from him. These messages can take many different forms, but their content and style are almost always a telling reflection of how the offender sees himself (or, particularly in the case of poison pen letters, herself).

In previous chapters, a number of cases have been described in which kidnappers and murderers have made contact, both written and verbal, either with the families of their victims or even with the police. At first, these messages may be simple and practical. However, if the case becomes long and drawn out and the police seem no closer to apprehending the criminal, they are likely to become taunting, as the criminal's confidence grows. Particularly interesting in this context is the case of Michael Sams.

THE SAMS LETTERS

Sams's first letter to the police in Leeds – a letter that later was found to have been

mailed when Julie Dart was already dead – announced that "a young prostitute has been kidnapped from the Chapeltown area last night and will only be released unharmed if the conditions below are met." After detailing the first stages by which the ransom money was to be handed over, the letter continued:

"The hostage will be fed and well looked after in a home rented for the purpose, she will be guarded 24 hours a day by P.I.R. [proximity infrared] detectors connected directly to the electricity mains. Once the monies have been withdrawn you will receive the address of the hostage. BEFORE ENTERING HOUSE THE ELECTRICITY MUST BE SWITCHED OF [sic] FROM OUTSIDE before opening the door or any movement will activate the detectors."

The letter had been typed on Sams's old Olivetti typewriter (later he used a word processor). It was a long letter, and toward the end the grammatical and spelling mistakes became more pronounced, as if the writer found it increasingly difficult to control his excitement. Even at

Profiler Paul Britton, known as the "jigsaw man" for his ability to construct offender profiles from disparate scraps of information and evidence.

this stage it was evident that he was playing a game with the police. For example, the mention of P.I.R. detectors and a demand that the female officer carrying the money should have "anti-tamper" and "transmitter detector" devices clearly visible in her car were unlikely to have a basis in reality. And Sams would know that if his victim were held in "a home rented for the purpose," the police would have little difficulty tracing him after her release.

Soon after Julie Dart's body had been discovered, a second letter arrived. Sams, having achieved his aim of securing intense police attention, was growing more confident. And as the increasing errors in the letter showed, more excited.

In his third letter, he discussed in detail the odds against any police operation resulting in his capture – an audacious demonstration of his belief in his own cleverness. Sams then went on to complain almost petulantly: "Julie was not bludgeoned to death (Jim Oldfield, *Daily Mirror*), she was rendered unconscious by three or four blows to the back of the head and then strangled. *She never felt a thing….*" This is not untypical of this type of murderer. They are quick to pick up on any inaccuracies in press reports and in this way try to justify their actions and build up their self-image.

The next letter followed two failed attempts to make telephone contact. It was handwritten and further indicated the killer's increasing confidence. It also contained eerie echoes of letters sent by Jack the Ripper. "Sorry no go last night was not free Monday afternoon for make-up to get hostage Monday evening – make-up takes hours," it began, and concluded, "Bye ring Tuesday." The police never revealed if they believed that Sams really intended to disguise himself with make-up, but it seems unlikely. A suggestion from Sams that he would use two people in a car in a "lover's lane" as intermediary hostages was also improbable and yet a further example of his game-playing.

By this time, one telephone contact had been made, but the resulting tape-recording was garbled, and only the word "services" was identifiable.

Following a further letter, the police were surprised to hear directly from Sams, who announced that his tape-recorder (used to disguise his voice) was broken. Unfortunately, they had not been expecting the call and his voice was not recorded. Then came a letter littered with errors in grammar and spelling that announced: "Game is now ababdoned [sic]."

Police issued this artist's impression of the man they believed killed Julie Dart. The drawing turned out to bear only a passing resemblance to Michael Sams.

The typed letter that later threatened to derail a train, however, was almost free of errors, suggesting that Sams had spent the previous two months in careful planning. The same was true of the letter that opened his campaign to kidnap an employee of a real estate agency, through which he made preliminary arrangements to view a house for sale. He telephoned the agency twice, but they had no reason to be suspicious. And the letter in which he announced the kidnapping of an employee, which the police believed he had written before the abduction of Stephanie Slater, was also almost free of errors. He was calm before the game and the chase began.

Sams was now so confident about his cleverness that he made open – as opposed to previously tape-recorded – telephone calls to the real estate agency. These calls were recorded and led to his eventual identification. A week after Stephanie was released, several copies of a letter from Sams were sent to the police and the media. Headed "The Facts," and beginning "I, being the kidnapper of Stephanie Slater," it claimed innocence of the murder of Julie Dart and the blackmail of British Rail. Most extraordinarily, it continued: "I am ashamed upset and thoroughly disgusted at my treatment of Stephanie and the suffering I must have caused to her parents.... Even now my eyes are all filled with tears, I wake up during the night actually crying, with a little luck Stephanie will get over it shortly. Myself? I do not think I ever will."

What is to be made of this letter, the last received before the arrest of Sams? Bob Taylor, the police superintendent in charge of the inquiry, considered it appalling, "full of self-gratification and pathetic pleas." Paul Britton's opinion, however, was that the offender realized that he was close to being identified and was "feigning contrition as if constructing a defense for himself."

He also thought it possible that during the week Sams spent with Stephanie as a captive he had in some odd way fallen in love with her and found it impossible to carry out his threats. The game had become too difficult to continue.

A police file picture of killer Michael Sams. Sams was sentenced to life imprisonment for murdering Julie Dart and kidnapping real estate agent Stephanie Slater.

This case also demonstrates the danger of reading too much into such letters. Britton, having correctly assessed that Sams was playing games with the police, concluded that the spelling and grammatical mistakes were deliberate, false trails that Sams was laying to divert the course of the investigation. It was only after his arrest that he was found to be slightly dyslexic.

THE CRIMINAL HAND

Graphology, the analysis of handwriting, can be approached in two different ways. Handwritten messages from offenders – whether demands for money, threats, or confessions of remorse – are almost always submitted to analysis by graphologists (the police prefer to call them "handwriting experts"). In this way, at some stage in the investigation, they can be compared with the writing of principal suspects. This analysis is essentially a physical, forensic technique – the identification and comparison of specific letterforms.

However, most practicing graphologists believe that they can detect predominant psychological characteristics in a person's handwriting. Some have claimed – with some justification – to be able to make an assessment that is as valuable as behavioral analysis or the indications of a typical MO.

Given the resistance of the police to what many people regard as hardly more scientific than astrology or palmistry, the study still has a long way to go before being accepted. Almost all the analyses

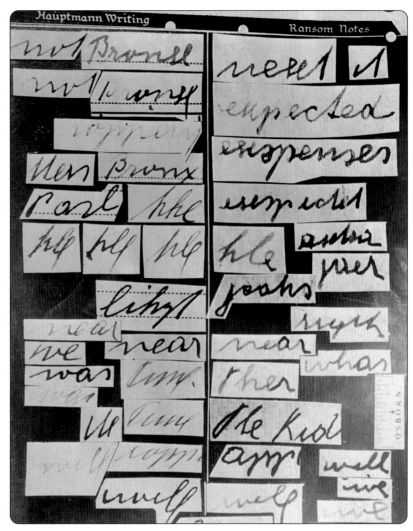

that have been carried out to date have been on the handwriting of apprehended criminals and made in hindsight. As Patricia Marne, the leading British graphologist and author of the book *The Criminal Hand* (1991), says, "No graphologist would describe a person as dishonest or criminal just on the strength of one or two of these give-away clues any more than a physician would diagnose a particular illness without taking more than one symptom into account."

In 1935, handwriting expert Albert S. Olson was called in during the trial of Bruno Hauptmann, accused of the murder of Charles Lindbergh's baby son. Olson revealed that the handwriting of the ransom notes closely matched the specimens of Hauptmann's writing.

ANTHRAX IN THE MAIL

Part of the text of one of the "anthrax letters," released by the FBI. They discounted, however, the possibility that the writer was an Islamic fundamentalist.

Following the anonymous mailing in the United States of letters laden with anthrax on September 18 and October 9, 2001, the FBI published a handwriting and behavioral profile of the writer on November 9. Stating that "it is highly probable, bordering on certainty, that all three letters were authored by the same person," they appealed for public help, hoping that someone might recognize the characteristics.

Although all the letters were written in capital letters, the first word of each sentence and the proper nouns were given a slightly larger letter; the FBI theorized that the writer was not practiced in lowercase lettering. The names and addresses on each envelope – all bought prestamped from a post office – had a noticeable downward slant from left to right. The writer wrote dates as "09-11-01," rather than "9/11/01," and used a formalized form of the figure 1. The word "cannot" was separated into two words: "can not."

The FBI release included a behavioral assessment. They stated that the offender:

"is likely an adult male;

"if employed, is likely to be in a position requiring little contact with the public or other employees. He may work in a laboratory.... He probably has a scientific background to some extent, or at least a strong interest in science;

"has exhibited an organized, rational thought process in furtherance of his criminal behavior;

"did not select the victims randomly.... The offender deliberately 'selected' NBC News, the *New York Post*, and the office of Senator Tom Daschle as the targeted victims (and possibly AMI in Florida). These targets are probably very important to the offender. They may have been the focus of previous expressions of contempt, which may have been communicated to others;

"lacks the personal skills necessary to confront others.... He may hold grudges for a long time, vowing that he will get even with 'them' one day. There are probably other, earlier, examples of this type of behavior;

"prefers being by himself more often than not. If he is involved in a

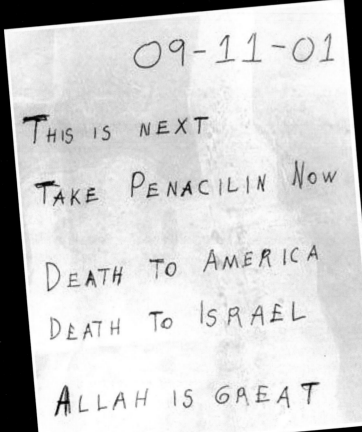

personal relationship it will likely be of a self-serving nature."

Finally, the FBI described the offender's probable before-and-after behavior. He might have displayed a passive disinterest in the events of September 11 – although these could have led him to become "mission-oriented." He would have become more secretive and exhibited an unusual pattern of activity, and he might have begun taking antibiotics unexpectedly.

At critical times – the mailing of the letters, the death of the first victim, media reports of each incident, and especially the deaths and illnesses of non-targeted victims – he might have exhibited significant behavioral changes. These could include an altered physical appearance; pronounced anxiety; atypical interest in the media; noticeable mood swings; an unusual level of preoccupation; unusual absenteeism; and altered sleeping and/or eating habits. To date, the anthrax mailer has not been identified.

Members of a biohazard team entering the Hart Senate Office Building on Capitol Hill, November 6, 2001. Many months passed before the building was declared free of anthrax contamination.

A BRIEF COURSE IN GRAPHOLOGY

Analysis and comparison of handwriting is a lengthy and complex business. Graphologists begin by dividing the writing into three zones. The middle zone is occupied by the vowels; the letters c, m, n, r, s, v, w, x, and z; the upper parts of g, p, q, and y; and the lower parts of b, d, h, and k. The upper and lower zones are occupied by the ascending or descending parts of the letters b, d, f, g, h, j, k, l, p, q, t, and y, and the corresponding portions of the capitals. Many young children are taught to write in copybooks, where these three zones are defined for them within four horizontal rulings. As their characters develop, however, most choose to express their individuality in deviations from the way they were taught to write. Graphologists believe that these deviations are strong indicators of personality.

The upper zone, they say, is the area of intellectual and spiritual qualities, ambition, and idealism. The middle zone represents the individual's likes, dislikes, rationality, and adaptability to everyday social life. The lower zone reveals the instinctive and subconscious urges together with the sexuality and materialistic interests of the writer.

These three zones are seldom of the same size. When they are, the appearance of the writing is likely to be relatively dull and stereotyped. A very formal hand – such as is taught in script classes – can present problems, as can handwriting that has been deliberately disguised. However, graphologists claim that no matter how much the letters differ from the "natural" hand of the writer, specific traits will be unconsciously betrayed. For instance, if a person has injured his or her right hand and is forced to take up writing with the left, he or she will gradually develop the same characteristics that were present in their original writing.

Two factors that are not definitively revealed are age and gender. Some people write a more mature hand at 20 than others do at 60. And everyone possesses male and female characteristics, both physical and psychological, in different proportions.

After considering the indications of the three zones, the graphologist goes on to look at a considerable number of individual indications. These cover the size and formation of the script, the apparent pressure and speed of the writing, the slant of the letters, whether the base line is level, uneven, or sloping, and the "form level." The form level is the visual quality of the writing, and it is divided into organization, spontaneity, originality, dynamism, harmony, and rhythm.

The size and regularity of the margins is important, as is the spacing between

> **The formation of the letters, both individually and in groups, is the most telltale element in graphological analysis.**

All handwriting occupies three zones. In copybooks these zones are strictly defined, but as handwriting develops writers express their individuality in deviations from this schoolroom style. Variations from the formal size of the zones are positive indications of the individual personality.

letters, words, and lines. Other indications are the formation of individual letters, particularly capitals, and especially the letter I; the crossbar of the t and the dotting of the i; loops, or their absence; punctuation marks; and numerals.

The signature, if there is one, is of great significance. A person's signature, although never identical from one given occasion to the next, remains fundamentally the same. On the other hand, two identical signatures would immediately arouse suspicion of forgery. If the style of the signature reveals significant differences from the body of the text, this is said to be evidence of a devious personality.

The formation of the letters, both individually and in groups, is the most tell-tale element in the identification and comparison of handwriting. The first aspect to be considered is the overall appearance of the text: whether the writing is angular, thread, arcade, or garland.

Angular writing is well controlled and definite. It indicates a shrewd and practical nature, inclined to be rigid in opinion, highly critical of others, and sometimes lacking a sense of humor.

Thread handwriting is the opposite of angular, the letters being stretched out and sometimes so illegible that their forms are difficult to analyze. It reveals an elusive and erratic personality who is likely to be a clever opportunist.

In arcade writing the tops of the letters *m* and *n* are particularly rounded. Graphologists believe that this reveals a secretive nature, inclined to be formal and uncommunicative about private matters. When *m* and *n* are unusually wide, with flattened tops, graphologists say that it is an indication of an amoral person who

is able to talk their way out of any situation.

Garland handwriting is easy and informal, reflecting its writer's nature. It is the opposite of arcade, the lower parts of the letters *h*, *n*, and *m* being rounded so much that there is often little or no distinction between *n* and *u*. The horizontal consistency of the handwriting – its adherence to a definable baseline – is also indicative. A straight line across the paper reveals self-discipline and good judgment.

A letter written from prison by a person serving a sentence for forgery. A good example of arcade handwriting, it suggests a secretive personality, while the wide formation of the letters "m" and "n" reveals the ability to be amoral while plausible.

Susan Smith killed her two young sons by pushing her car – with them in the backseat – into a lake. Her handwriting is practical, but immature. The varying size of individual letters suggests a reluctance to face up to her actions.

A slope upward is said to indicate ambition and optimism while a downward slope reveals pessimism and depression. And a stepped baseline, going up and down irregularly, can be an indication of mental confusion.

Next are the individual letters. Capitals are particularly revealing, as they indicate the individual's estimation of their own importance. "Enrolled" capitals, in which the ends of the strokes are scrolled in on themselves, are held to be a sure sign of deceitfulness.

One of the most revealing of all is the capital *I*, because it directly represents the ego. A small *I*, for instance, can suggest a lack of self-confidence, while one that is flamboyant and exaggerated is a sure sign of someone who wants to be the center of attention. An *I* that is a simple downward stroke suggests a personality that is self-assured, intelligent, and well-balanced. An *I* that swings or leans to the left, whatever its form, is said to indicate an inability to enjoy life, possible guilt about some past event, and a propensity to deceive.

Two other letters are also particularly significant because they involve breaking the flow of the handwriting. These are *i* and *t*. The position and form of the *i* dot are not only easily identified charact-eristics, but according to graphologists they reveal a lot about the personality of the writer. When the dot is placed to the left of the stem of the *i*, it denotes caution and hesitation when the writer is faced with making decisions. Placed to the right, it indicates forward thinking and practicality. Exactly above it is a sign of an eye for detail but not much creative imagination. And when the dot is connected to the following letter it indicates intelligence and the ability to adapt and plan ahead.

A heavy dot is frequently a sign of pessimism or depression; conversely, a very lightly drawn dot can indicate lack of interest or a low level of vitality. When the dot appears as a small, elongated line, it is a sign of an oversensitive nature, difficult to please. A very large dot racing to the right reveals an impatient disposition, frequently forced to make rapid decisions.

When I left my home on Tuesday, October 25, I was very emotionally distraught. I didn't want to live anymore! I felt like things could never get any worse. When I left home, I was going to ride around a little while and then go to my mom's. As I rode and ride and rode, I felt even more anxiety coming upon me about not wanting to live. I felt I couldn't be a good mom anymore but I didn't want my children to grow up without a mom. I felt I had to end our lives to protect us all from any grief or harm ~~██████████~~ I had never felt so lonely and so sad in my entire life. I was in love with someone very much, but he didn't love me, and never would. I had a very difficult time accepting that. But I had hurt him very much and I could see why he could never love me. When I was @ ~~John D. Long Lake~~, ~~██████~~ I had never felt so scared and unsure as I did then. I wanted to end my life so bad and was in my car ready to go down that ramp into the water and I did go part way, but I stopped. I went again and stopped. I then got out of the car and ~~███~~ stood by the car a nervous wreck. Why was I feeling this way? Why was everything so bad in my life? I had no answers to these questions. I dropped to the lowest when I allowed my children to go down that ramp into the water without me. I took off running and screaming "Oh God! Oh God, NO!" What

A dot in the form of an arc or a horizontal line indicates creativity and a powerful imagination. A sharp dot like an arrow is a sign of anger and bitterness. And when the dot is drawn as a circle – most often found in the handwriting of young women who want to be seen as different from their acquaintances – it is a sign of an unusual sense of humor and of a desire to express the writer's self-importance.

The bar of the letter *t* is equally revealing. Connected to the following letter, it is an indication of speed and therefore of a mind that is quick thinking. And when it races away to the right it reveals energy and ambition. But placed to the left it betrays a nature that tends to be introverted and cautious. A bar sloping downward can be a sign of suppressed anger, but an upward slant suggests optimism and ambition.

A bar drawn with care in the middle of the stem is a sign of a cautious and methodical nature; low on the stem, it suggests depression; and, at the very top, it is an indication of an impractical, daydreaming nature.

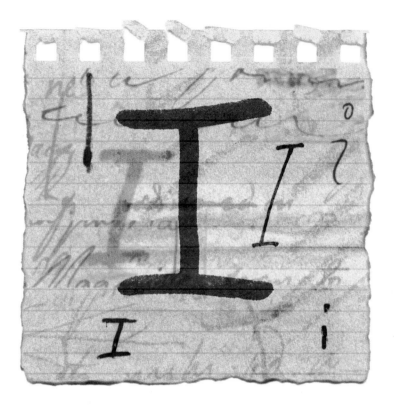

Handwriting experts consider the letter "i" to directly represent the ego.

If the bar is long, drawn over a large part, or all, of the word, it can be a sign of protectiveness toward family and friends, but equally can represent a patronizing attitude to those outside the immediate circle. A small hook at either end indicates tenacity, but a triangular bar moving back against the direction of the writing is a sign both of aggression and of sexual disappointment. Finally, there is the bar formed as an upward continuation of the down stroke, which graphologists regard as a sure sign of the habitual liar.

These, then, are some of the more important factors that graphologists look for when examining handwriting. They serve not only to confirm if two specimens are by the same hand, but they also provide a degree of insight into the personality of the writer.

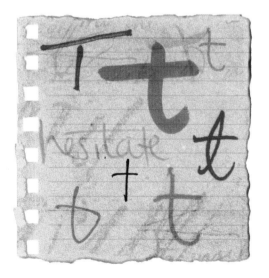

Left: The way the letter "t" is written can reveal a lot about the writer's personality. A bar drawn with care indicates a cautious and methodical nature.

The face of vicious serial killer Ted Bundy. Handsome and charming, and with his arm in a fake plaster cast, he attracted his early victims by asking them to assist him.

CASE STUDY: TED BUNDY

Although he came from a poor family, Theodore Robert Bundy was well educated. He attended Stanford University on a scholarship in Chinese studies and graduated with a B.S. in psychology. Ironically – in light of his subsequent career as a serial killer – he found employment as a psychology assistant to the Seattle crime commission. Good-looking, he was charming and witty with women – but he had an uncontrollable urge to kill them.

During the first eight months of 1974, female college students began to disappear in the Seattle area at a rate of about one a month. Then the disappearances suddenly ceased. Bundy had moved to Salt Lake

City, where he entered the University of Utah to study law. Shortly after, co-eds began to disappear in Utah and in neighboring Colorado. In August 1975, Bundy was arrested in Salt Lake City for driving in a suspicious manner, and while he was in detention a hair was discovered in his car that matched that of a girl whose raped and battered body had been found in Colorado.

Bundy was extradited to Aspen, and imprisoned to await trial for murder. He charmed his warders and the prosecutors alike, and was allowed to visit the law library to prepare his own defense. There he managed to open a window, drop 20 feet to the ground, and escape. He was recaptured eight days later, hiding in a deserted mountain cabin, but in December 1977 he dug a hole through the ceiling of his prison cell and escaped again.

Within a few days, Bundy was in Tallahassee posing as a graduate student at Florida State University. On January 15, 1978, he broke into the Chi Omega sorority house and ran from room to room, attacking and raping the women, killing two with a wooden club and leaving two others badly injured. On the way back to his lodging, he raped and killed another student in her bed.

Bundy might never have been suspected if he had left Florida at that time. However, on February 9, he kidnapped a 12-year-old girl, raped and killed her, and hid her body in an empty shed. On the night of February 14 he was arrested – once again for driving in a suspicious manner. And it was only then that he was identified.

When he was brought to trial, the most telling evidence against him proved to be a bite mark on the buttock of one of the victims of the Chi Omega attack, which matched that of Bundy's teeth.

Using his knowledge of the law to lodge a succession of appeals, he remained on Death Row for 10 years, but was eventually executed in January 1989. In the end, he admitted to 23 murders, although at least 15 more have been attributed to him.

After his arrest, Ted Bundy told a detective: "Sometimes I feel like a vampire." He was finally identified by his bitemark on the buttocks of one of his last victims.

BUNDY'S HAND

In her book *Handwriting of the Famous and Infamous* (2001), leading graphologist Sheila Lowe – British-born, but now a court-qualified handwriting analysis expert in California – gives a summarized analysis of Bundy's script. What is immediately striking is the long hooked initial strokes, called "harpoons," that begin nearly every word. They are particularly prominent in the left-hand margin, which, says Lowe, denotes anger and resentment concerning events in the past. "The slack, disturbed rhythm, which is often a feature of the handwritings of criminals," she writes, "signifies a poorly integrated personality. The crowded spatial arrangement interferes with his ability to keep a clear perspective and understand what is appropriate."

In the middle zone, which represents the subject's likes, dislikes, and social relations, Sheila Lowe points out the variation in zone widths and word spacing. This, she says, indicates inconsistency in his interactions with others. "The harpoons, coming from the lower zone where they don't belong, suggest holding on to past slights, real or imagined. They are hidden at first, and then shoot up angrily into the middle zone."

The lower zone is associated with sexuality. Here the unusually deep strokes, which penetrate in many cases into the upper zone of the succeeding line, indicate strong drives. Yet at the same time the rhythm of movement is slack and, says Lowe, "the tremulous strokes reveal Bundy's discomfort in this area."

In contrast, the upper zone – with the exception of the overblown *I* – is relatively confined, denoting a narrow-minded attitude, "with no room for anything new or different." The personal pronoun "has a heavy emphasis on the 'mother' area of the upper loop, so his view of women was unrealistic and had no 'father' figure to balance it."

She concludes: "The crowded picture of space, large personal pronoun, and rigid initial strokes suggest a need for power and control. There is not the kind of elastic rhythm that attests to positive energy and willingness to work hard for what he wanted, so he would take control in whatever was the easiest way for him at the moment."

A portion of a letter written by serial killer Ted Bundy. The large personal pronoun, crowded writing style, and sharp initial strokes suggest a desire for power and control.

CASE STUDY: ARTHUR SHAWCROSS

In 1972, 26-year-old Arthur Shawcross murdered two young children, a girl and a boy, in Watertown, New York. Because the medical evidence on the little girl, who had been raped, was badly botched, and the boy's remains were discovered six months after his death, there was a possibility that a trial for murder would fail. The DA therefore made a plea bargain. The charge was reduced to manslaughter in exchange for a confession, and Shawcross was sentenced to 25 years in prison.

He was released on parole in 1987 after 14 years and settled in Rochester, New York. Corpulent and gray-haired after his time in prison, he looked as if he no longer posed a threat to anyone. However, between March 1988 and January 1990, he sexually assaulted, killed, and brutally mutilated at least 11 local prostitutes. It was not until a police helicopter spotted him returning to the scene of his latest

Over 40 years old, corpulent and gray-haired, Arthur Shawcross seemed to be an unlikely serial murderer. But over a period of less than two years he murdered and seriously mutilated at least 11 prostitutes in Rochester, New York.

Shawcross was found to have the XYY chromosome; in this letter from prison he discussed the theory that his abnormality was associated with a criminal personality.

murder that Shawcross was arrested and eventually confessed. At his trial he was judged sane, found guilty on 10 counts of murder, and sentenced to 125 years in prison.

Psychiatrist Richard Kraus made a detailed study of the case. Shawcross was found to carry the XYY chromosome, a hereditary abnormality that has been connected with violent behavior. His urine also showed a very high level of a chemical metabolite called cryptopyrrole, usually absent in humans. Kraus theorized that the combination of these two factors made Shawcross a "walking time bomb," unable to control his rage and liable to relieve his emotions in violent ways.

Shawcross's note from prison, commenting on the XYY chromosome, is a mass of graphological contradictions. The difference between the text and the signature is striking and indicates a repressed and devious personality.

The note itself is written largely in capital letters, although with consistent exceptions for *b*, *g*, and *i*. This shows that Shawcross wanted to be sure that he would be clearly understood, fearing that his normal handwriting would be insufficiently legible. In his case, it is also a sign of emotional immaturity and semi-literacy – yet at the same time he reveals a sophisticated vocabulary, mostly correctly spelled, and evidence of considerable reading. The wide word and letter spacing symbolizes the amount of personal space the individual demands, revealing that Shawcross was not willing to form close social relationships.

Sheila Lowe has pointed out that an important aspect of Shawcross's writing is its rigidity. This is characteristic of many criminals in prison, because they are forced to keep their violent impulses under control. The way in which certain letters are suddenly written larger or jump up from the line, she says, suggests impulsive and antisocial behavior. Even the forward slant of the letters, often taken as a sign of inner anger, is inconsistent and yet another sign of an erratic, unreliable personality.

The deep descent of the looped *g* into

XYY — FACT OR MYTH

Most people were convinced that, without exception, the motivation's behind the antisocial acts of violent people had their roots in poverty, broken homes, childhood traumas, parental neglect, or parental ineptitude. Sociologist insisted criminals were spawned primarily by social systems founded on injustice. Most psychologist appeared secure in belief that any neurosis or psychosis could be explained in terms of Freudian or Jungian theroy. But was it possible that some people were rotton from the start, hopelessly corrupt before environment had an opportunity to effect them? Was that a reactionary, Medieval thought? It's been said a great deal of a man with the XYY — i's a genetically ordained criminal type that inspired so much scientific research over the past few years. Could it be some people can be born less human or civilized then others? Can it be caused by chemical or genetically reasons that no one yet has thought seriously about yet? This can be misinterpreted! If I am a genetic or chemical inferiority, then lets look into every aspect of others who may have what I do.

Was I born to do evil or trained?...

I am a special creature, born more or less than human...

Arthur J. Shawcross

ADOLF HITLER

Sheila Lowe analyzed the handwriting of Adolf Hitler, dictator of Germany from 1933 to 1945. She states that the combination of angles and thread connecting forms indicates his cruel and angry nature. "Add to that the muddy pressure, for the hallmarks of an inharmonious personality with an authoritarian attitude and lack of tolerance."

Like his signature, Hitler's handwriting tends to slope downward from left to right. This is a sign of depression and lack of optimism; "there was little or no sense of humor to relieve the irritability," she wrote. The formation of the letters is narrow, indicating shyness and inhibition. By contrast, the tall capital letters reveal Hitler's pride in his achievements, while the heavy *i* dots and horizontal strokes in the upper zone suggest an explosive temper.

The strong pressure and horizontal expansion indicates that, at the time this note was written – when the Nazis were dominant throughout Europe – Hitler was still a man of impulsive action and decision with almost limitless stamina and endurance. The linear nature of the writing shows a strong intellectual tendency, but the lack of development of the upper zone suggests that Hitler was rigidly confirmed in his personal beliefs and intolerant of differing opinions.

The handwriting of Adolf Hitler reveals an authoritarian attitude and lack of tolerance.

the lower zone and the way in which the loop itself is small and cramped is a sign of sexual frustration.

Sheila Lowe suggests that a lack of nurturing in the early stages of Shawcross's life is revealed in the movement of the loop far to the left. The strongly drawn capital *I*'s show that Shawcross was a man of fiercely held opinions, which he was likely to maintain with violence. This fact was substantiated by evidence given by his fourth wife. His inner anger is also revealed in the strongly placed dot of the *i*.

CASE STUDY: A SWISS BOMBER

Few graphologists would attempt to construct a profile from no more than a written name and address. However, in 1962, the leading Swiss graphologist in Zurich, M. Litsenow, agreed to try and achieved a remarkable success.

In June and July of that year, five bombs exploded in Lucerne, three of them in the elevators of prominent restaurants, resulting in the injury of five people and 100,000 Swiss francs worth of damage. Police inquiries into the source of the detonators led to a legal firearms dealer,

whose register was signed with the name "Afled Späni" and an address that proved to be false.

Consulted by the police, M. Litsenow declared that the signature was also obviously false because it lacked ease and spontaneity – and was probably a failed attempt to write "Alfred." The writer, he said, was a male of only average intelligence, a poor student whose age was at the younger end of the 20–40 bracket. He was not a professional technician, and his unstable personality and sense of inferiority indicated that he was unlikely to have regular contact with the public. The graphologist suggested that he was a casual, nonskilled worker such as a warehouseman. He would also be conventional in appearance and reasonably athletic and fit.

M. Litsenow, in view of what he considered the bomber's feelings of inferiority, said that his motive was a desire to feel important. His parents were possibly alcoholics or had separated. He might already have been in trouble with the authorities over a minor incident, and the graphologist suggested that police consult the files of the social services agencies.

On the basis of M. Litsenow's recommendations, the police questioned half a dozen suspects and eliminated all but one. He was Anton Fähndrich, a 20-year-old warehouseman. Conventionally dressed, he lived in a church hostel – and had recently won two boxing championships. Comparison of his handwriting with the false register entry revealed many typical characteristics, and during questioning he declared that his parents had separated after being arrested several times for drunkenness. At first, he denied responsibility for the bombings, but eventually confessed that they were from "a need to revenge himself on society."

AUTHOR ATTRIBUTION

Another approach to written communications, almost the opposite of graphology, is textual analysis, the detection of specific ways of expression, phrases, and even individual words.

Investigators can use textual analysis to help identify a suspect. This can be very important in a case that has become long and drawn out. If several messages have been received from the presumed offender there is the danger that "copycat" communications – whether from hoaxers or from disturbed individuals – can divert the course of the investigation if they are not discounted.

Many academics have been involved in developing this type of textual analysis. One of the most famous in recent years is

The novel **Primary Colors** *by Anonymous was made into a movie in 1998, starring John Travolta as the presidential candidate. Professor Don Foster identified the author of the novel, journalist Joe Klein, by analyzing the vocabulary of his writing style.*

Journalist Joe Klein (left, with his publisher Harold Evans from Random House) confesses to his authorship of Primary Colors *before a crowded press conference in July 1996.*

Don Foster, an American academic who achieved sudden fame in 1996 when he correctly identified the author of the novel *Primary Colors* (by "Anonymous") as the *Newsweek* journalist Joe Klein. The book was a satirical but sympathetic fictional-ization of presidential candidate Bill Clinton's 1992 campaign. It rapidly became a bestseller, and readers were clamoring to know who the author was.

Foster, a professor at Vassar, had made the front page of the *New York Times* two weeks earlier when he identified the

author of a poem – published in 1612 as "A Funeral Elegye" – as William Shakespeare. Now he was challenged by *New York* magazine to name the writer of *Primary Colors*.

Foster had spent 12 years researching the authorship of the "Elegye," analyzing its vocabulary and structure. At the same time, he had enjoyed correctly identifying anonymous critics of his manuscript about the author of the "Elegye." Provided with the writings of 35 journalists suspected of writing *Primary Colors*, he set his computer to work, and after many searches for unusual words he found a wealth of adverbs and adjectives, as well as strange compound nouns, that were typical of columnist Joe Klein.

Klein, naturally, denied authorship but an article in the *Washington Post* in July 1996 revealed that handwritten notations on the original manuscript of the book exactly matched Klein's handwriting. At a press conference held soon after by his publishers, Klein appeared wearing a false nose and moustache. Removing the disguise, he announced: "My name is Joe Klein, and I wrote *Primary Colors*."

ANOTHER MAD BOMBER

A famous case from the 1990s illustrated the importance of textual analysis. For 17 years – from 1978 through 1995 – the FBI was engaged in a hunt for a man who, because he revealed a particular antipathy to university scientists, became known as the "Unabomber." At first, the bombs that he manufactured were little more than booby-trap devices, but they later became increasingly dangerous, causing the deaths of three people and injuring 29 others.

Between 1987 and 1993, there were no bombings, and the FBI began to suppose that the offender was in prison or a mental institution, or even dead. Then in June 1993, he renewed his campaign, referring to himself as "FC." This was a disturbing reflection of George Metesky's signature of "FP." At a loss, the FBI drew up a list of 50,000 "possible" and 600 "plausible" suspects – even

Professor Don Foster's name was among them. The manhunt grew more urgent in December 1994, when the Unabomber's most powerful device to date decapitated a New Jersey advertising executive as he opened a package addressed to his home. In April 1995, a similar bomb killed a lobbyist for the logging industry in California.

In June 1995, a 35,000-word manifesto entitled "The Industrial Society and its Future" by "FC" was sent to the *New York Times* and the *Washington Post*. A report on the document was read by an American professor of philosophy, Linda Patrik, who was on vacation in France – and who happened to be married to Ted Kaczynski's brother David. The FBI was said to be searching for a native of Chicago with connections to Salt Lake City and Northern California; a loner with experience in woodworking and knowledge of explosives, and yet with a deep-seated resentment of modern technology.

Professor Patrik was struck by the apparent similarities with the brother-in-law she had never met, and when her husband joined her in Paris she shared her suspicions with him. After the couple returned to the United States, David was able to compare the text of the manifesto with the many angry letters Ted had sent him from his lonely cabin in Montana. Reluctantly, he agreed with his wife: the Unabomber was his brother. David Kaczynski contacted the police. When FBI agents raided the cabin and arrested Kaczynski, they found ample

Professor Patrik was struck by similarities with the brother-in-law she had never met and shared her suspicions with her husband.

On June 23, 1993,
Dr. David Gelernter,
associate professor of
computer science at Yale
University, received a
package mailed by
the Unabomber. It
exploded, ripping off
his right hand and
destroying the sight of
one eye and the hearing
in one ear.

evidence of his bomb-making activities and many documents, including a long "Autobiography," in which he expressed all his angry feelings.

All along, the FBI had been following a false trail. A profiler had concluded that the Unabomber was born between 1957 and 1962, whereas Kaczynski's date of birth was 1942. In a hunt that eventually cost more than $50 million, school yearbooks and university employee and student directories were cross-checked, but

only as far back as 1970. Kaczynski, who graduated from high school in 1958, had taught at the University of Michigan and the University of California at Berkeley before giving up his academic career – to become a lonely hermit – in 1969.

Foster's work in the *Primary Colors* case had attracted the attention of the FBI, and he was subsequently invited to advise the prosecution on the acceptability, as evidence, of using Ted Kaczynski's writings to prove he was the "Unabomber."

Foster's analysis of the Unabomber's texts was, of course, colored by hindsight. However, he correctly detected evidence of Kaczynski's favorite reading matter and suggested that this would have made it possible for the FBI to identify the research libraries in Northern California Kaczynski had visited and even the days on which he had handled specific books.

Foster's subsequent advising of the FBI in other cases cannot, he says, be made public. And sadly, he has retracted his identification of Shakespeare as the writer of the "Elegye."

PSYCHOLINGUISTICS

As formal scientists, psycholinguists study how people learn to communicate with one another, how the developing child shows an innate understanding of the rules of sentence construction, and how language is structured. In recent years the FBI, among others, has shown an interest in a branch of this subject, theorizing that an analysis of

an UNSUB's (unknown or unidentified subject) use of language – as distinct from textual analysis – could contribute to his behavioral profile.

As a paper in a recent FBI *Law Enforcement Bulletin* put it: "One type of behavior often overlooked, or underused, exists in the offender's actual language. The offender's written or spoken language can provide investigators with a wealth of information.… Both have features that may reveal an individual's geographical origins; ethnicity or race; age; sex; and occupation, education level, and religious orientation or background."

As an example, the paper points out how even very minor differences in location can result in detectable differences in vocabulary: "In Pennsylvania, when people from Philadelphia want a carbonated soft drink, they tend to ask for a 'soda,' whereas those from Pittsburgh more likely request a 'pop'."

Research has shown that men and

FALSE ACCUSATIONS

The FBI was asked for advice in the case of a woman who had received seven threatening letters (the latest of which suggested rape), and named two men who might be responsible. Behavioral Science analysts at first concentrated on which of the two males revealed the more likely behavioral characteristics, but soon realized that the writer of the letters was most probably female. They asked the police investigators if the woman had recently experienced a major stressful event, and learned that she had been dating a local police officer, who had returned to his wife just four days before the letters started arriving. When he learned of the threatening letters, he again left his wife and

renewed his relationship with the "victim." The motivation behind the writing of the letters was obvious, and the FBI analysis cleared both suspects.

In addition, psychiatrists and psychologists have recently identified features of language that can be associated with characteristics of personality such as impulsiveness, anxiety, depression, paranoia, or the desire for power and control. The FBI is currently carrying out research on the relationship between the language used in threatening communications and the potential risk of actual violence. Other research is directed at the examination of suicide notes to determine if the suicides are actually disguised homicides.

women tend to have slightly different patterns of language. Female writers are more likely to use tentative expressions, such as "it seems like," or "I suppose I should have." Their language also expresses emotions more freely, as in "I felt compelled" or "I hope," and makes greater use of words like "so" and "such": "It was so disgusting," "I felt such pleasure."

CASE STUDY: A VITAL NOTE

Shortly before 6:00 A.M. on the morning of December 26, 1996, Patsy Ramsey made a 911 call to the police in Boulder, Colorado, to say that her six-year-old daughter JonBenét had been kidnapped. She said she had found a ransom note at the foot of the stairs. It consisted of three sheets of paper, claimed to be from "a small foreign faction," demanded $118,000, and was signed "S.B.T.C." Although the note threatened that, "Speaking to anyone about your situation…will result in your daughter being beheaded," Mrs. Ramsey called not only the police but also several close friends and a local minister, who all arrived soon after the two police officers who were sent to the home.

The policemen made a brief search of the house. There were no signs of a forced entry and no footprints other than their own and those of the visitors to the family in the surrounding snow. The note stated that a telephone call would be made between 8:00 and 10:00 A.M., with instructions on how the ransom money was to be delivered. When no call had been

> "Give me two lines of a man's handwriting," said Cardinal Richelieu, 17th-century chief minister of France, "and I will hang him."

received by 1:00 P.M., the detective in charge ordered a thorough search of the premises. JonBenét's father, John Ramsey, immediately went to the basement wine cellar and emerged a few minutes later carrying the body of his daughter.

When a murder occurs in a home with no sign of an intruder, police suspicion naturally focuses upon the parents. Suspicion was heightened when police discovered that the ransom note had been written on sheets from a notepad in the house and with a red felt-tipped pen that was also found. They also found traces of a "practice note" similar to the one Patsy Ramsey handed to the police.

The Ramseys have steadfastly maintained their innocence. A wealthy couple, they have been able retain the services of top attorneys and forensic experts. A grand jury that sat for nearly a year in Boulder was unable to reach a conclusion. Nonetheless, Colorado Governor Bill Evans described the Ramseys as being "under the umbrella of suspicion."

Naturally, the "ransom note" has become the subject of intense scrutiny. Don Foster at first offered to assist John and Patsy Ramsey. Later, in January 1997, he was consulted by the Boulder police. In his book *JonBenét: Inside the Ramsey Murder Investigation* (2000), former Boulder detective Steve Thomas reported:

"Foster dissected the ransom note, explained that the writing contained intelligent and sometimes clever usage of language, and said the text suggested

Far right: The Unabomber, solitary hermit Ted Kaczynski, was finally apprehended at his isolated cabin in Montana, following the identification of his style of writing by his brother David and David's wife, Linda Patrik.

The murder of six-year-old beauty queen JonBenét Ramsey remains unsolved. Her body was discovered in the basement wine cellar of the family home by her father, after a preliminary police search had revealed nothing.

someone who was trying to deceive.

"The documents he studied from Patsy Ramsey, in his opinion, formed 'a precise and unequivocal match' with the note that included such things as her penchant for inventing private acronyms, spelling habits, indentation, alliterative phrasing, metaphors, grammar, vocabulary, frequent use of exclamation points, and even the format of the handwriting on the page."

Psycholinguist Dr. Andrew G. Hodges goes even further in his book *Who Will Speak for JonBenét?* (2000), analyzing what he describes as "thoughtprints." He maintains that every action we take has an underlying motive that is evident to the subconscious and emerges in a detectable way in everyday communication. "I look at each word for two meanings, not one," he says. The second meaning is the subconsciously encoded message. Hodges and two colleagues submitted a 70-page analysis of the Ramsey ransom note to the Boulder District Attorney. In it they assert that a killer cannot avoid confessing in some manner, and that the "ransom note" reveals:

- the killer is a woman
- whoever wrote the note participated in the murder
- her husband participated in the murder and cover-up
- she expects to be caught
- her motive was anger and deep pain
- she offers details about what precipitated the murder

One page from the three-sheet ransom note, signed "S.B.T.C.," that JonBenét's mother, Patsy Ramsey, reported that she had found at the foot of the stairs. "Speaking to anyone about your situation, such as Police, F.B.I., etc.," it reads, "will result in your daughter being beheaded."

- the note itself was prompted by psychological motives
- the ransom amount indicates that this was not really a kidnapping
- the victim was dead before the note was written
- the note is a story told by a firsthand witness and whoever finds it should "listen carefully" – this is repeated four times.

In his book *Author Unknown* (2000), Don Foster avoids any explicit attempt to assess the psychological makeup of the writers he analyzes. Nevertheless, he remarks that the spelling in the O. J. Simpson "suicide note" could be taken as significant. Where Simpson presumably intended to write that it was "tough splitting up" with his wife Nicole but that they had "mutually agreed" that the separation was necessary, he actually wrote that it was "tough spitting," and the breakup was "murtually agresd."

Some psycholinguists might argue that "spitting," and the echoes of "murder" and "aggression" reflected the subconscious concerns of the writer.

The FBI is now applying the principles of psycholinguistics to Internet and computer crime. Although the World Wide Web seems to provide an opportunity for complete anonymity, researchers have discovered that language even in the form

John and Patsy Ramsey, JonBenét's parents, protesting their innocence on CNN's "Burden of Proof."

of computer codes can provide clues that can lead to the identification of hackers and other dangerous offenders.

THE TELLTALE VOICE

The identification of a recorded voice is highly subjective. We all believe we can recognize the voice of a family member or friend on the telephone, but experience shows that mistakes are quite often made. Distortion can take place on the line and – if the message is tape-recorded – technical shortcomings in the equipment can make the voice unrecognizable. And when the recorded voice of a criminal is broadcast

on radio or television in the hope that someone will recognize it, people often name several individuals, convinced that they "know the voice." Police also need to be able to distinguish between the genuine recorded voice of an offender and that of hoaxers and disturbed people who confess to the crime.

Experts in linguistics can detect the subtleties of regional accents and characteristic nuances of sentence structure, and can claim to be able to place a person's voice within quite narrow geographical or social limits. The hunt for the Yorkshire Ripper, for example, was diverted for many weeks after experts had – no doubt correctly – pinpointed the voice on the "Ripper tape" to a small group of villages in northeast England. The tape proved to be a false lead recorded by an unidentified hoaxer.

However, even correct recognition of someone's voice, which can lead the police to the offender, is seldom acceptable as

West Yorkshire's Assistant Chief Constable, George Oldfield (center), listens to an eerie tape recording of the man erroneously thought to be the Yorkshire Ripper.

A LANDMARK CASE

During the 1965 riots in the Watts district of Los Angeles, a CBS television interviewer recorded a conversation with a young man. The man boasted of his fire-raising activities, but kept his back to the camera. Later the police arrested 18-year-old Edward Lee King on suspicion, and asked Lawrence Kersta – who had developed the voiceprint technology – to compare the TV recording with tapes of King's voice.

At a long, drawn-out trial, Kersta testified that the voiceprints were identical, and King was convicted of arson. He appealed on the grounds that providing a recording of his voice amounted to self-incrimination. Eventually the U.S. Supreme Court ruled that the "right of privilege" against self-incrimination did not apply in such cases.

A pall of smoke rises over the Watts district of downtown Los Angeles during the riots of August 18, 1965.

evidence. Forensic scientists were delighted, therefore, when the "voiceprint" technique of speech identification – one that could be demonstrated practically to a jury – was accepted by a United States court in 1967.

KERSTA'S SPECTROGRAPH

During World War II, U.S. soldiers eavesdropping on German military radio communications needed to be able to distinguish and identify different speakers. They asked scientists and engineers at the Bell Telephone laboratories in New Jersey to look into the possibility of developing an electronic method of recording and identifying specific voice patterns. One of those assigned to the problem was Lawrence Kersta, who continued his research after the war. In 1963, he finally perfected a method of recording the pitch, volume, resonance, and articulation of a human voice, in what he called a "spectrogram."

The sound of the human voice depends upon two general factors. One is the

Comparing voiceprints on a computer screen. the two principal displays are the characteristic waveform outputs from the electronic frequency filter, and the plot between them is the comparison spectrogram.

When Clifford Irving forged the memoirs of Howard Hughes, the eccentric millionaire broke 15 years of silence to make a two-hour telephone call from his hideaway in the Bahamas to denounce the work as "totally fantastic fiction." Lawrence Kersta examined recordings of the conversation and compared them with a recording of a speech Hughes had made more than 30 years earlier. He concluded that the voice on the telephone was undoubtedly that of Hughes.

resonance that is developed in the throat and chest and in the rounded cavity of the mouth by the movements of the larynx and vocal cords. The other, more immediately distinctive, is the use of the articulators – lips, teeth, tongue, jaw, and so on – which define the sounds of individual words. As Kersta wrote: "The chance that two individuals would have the same dynamic use of patterns for their articulators would be remote. The claim for voice pattern uniqueness, then, rests on the improbability that two speakers would have the same vocal cavity dimension, and articulator-use patterns, nearly identical enough to confound voiceprint identification methods."

Kersta's spectrograph is comprised of four parts: a high-quality magnetic tape recorder; a frequency filter; a tape scanning drum; and an electronic stylus that records the readings on sensitive

paper. The output can also be displayed on screen or recorded on computer for more detailed analysis.

A 2.5-second specimen of speech, recorded on the tape, is fed over and over again through the electronic frequency filter. This progressively selects narrow bands of frequencies, moving gradually from low to high, and the stylus records their relative intensity. The final print is a pattern of closely spaced lines representing all the frequencies of the person's voice and their intensities. Two types of voiceprint can be obtained.

One, the bar print, is what is usually produced as evidence in court. The horizontal scale represents the length of time of the recording, the vertical scale represents the frequencies, and loudness is represented by the density of the print. The second type is the contour print, which displays the more complex characteristics

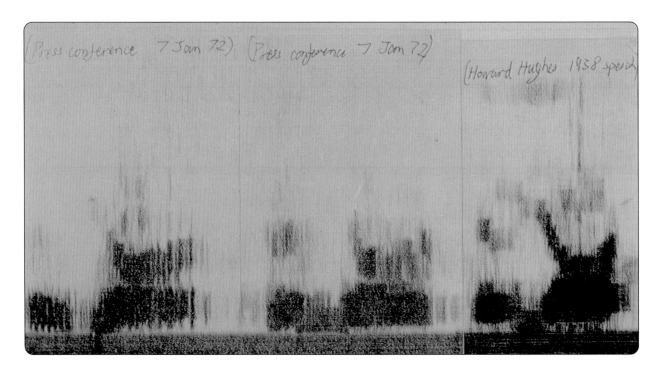

THE FATAL POACHER

Following the 1967 acceptance of voiceprints in the case of Edward Lee King, a number of police authorities in the United States took up voice identification. The Michigan State Police were the first to do so.

In September 1971, a game warden in the region of Green Bay, Wisconsin, was reported missing, and the following day his decapitated body, shot several times, was found in a shallow grave with his head buried nearby.

The police officer in charge of the investigation, Sgt. Marvin Gerlikovski, reasoned that the killing had been in revenge, and required all those who had previously been arrested by the warden to be interrogated.

Those who could not provide a firm alibi were asked to take a polygraph ("lie detector") test.

Only one refused: Brian Hussong, a man well known locally as a poacher.

Gerlikovski obtained a court order enabling him to place a wiretap on Hussong's home telephone and recorded his conversations. One was with Hussong's 83-year-old grandmother, who assured him that his guns were well hidden. A search of her home revealed the guns and examination by a ballistics expert confirmed that one had killed the game warden.

At Hussong's trial, his grandmother denied all knowledge of the weapons. However, an analyst from the Michigan Voice Identification Unit produced voiceprints in evidence, and demonstrated the identity of the grandmother's voice, which clearly differed from that of Hussong's other relatives.

---

of the voice and is more suitable for computer analysis.

During his research to establish the unique quality of the voiceprint, Kersta made 50,000 recordings of individual voices. Although many sounded similar to the ear, the spectrograph clearly demonstrated their differences. Kersta even made use of professional mimics and showed that while their imitations were very like the real thing, their voiceprints were distinguishably different.

More than 5,000 cases of voice identification have been handled by certified voiceprint examiners in the United States to date. The cases covered include murder, rape, extortion, drug smuggling, burglary, bomb threats, terrorist and organized crime activities, political corruption, and tax evasion. An FBI survey of its own performance in 2,000 cases reported an error rate of only 0.31 percent in false identifications and only 0.53 percent for incorrect eliminations.

The presentation of voiceprints in evidence, as with fingerprints, requires the establishment of an agreed number of identification points. At the moment there is no common standard. The U.S. Internal Revenue Service demands a minimum of 20 identical speech characteristics, but other agencies require only 10 or more. So far, other countries have remained doubtful of the evidential value of voiceprints, and they have been slow to adopt this revolutionary technique.

Obviously, voiceprints are of little use until they can be compared with that of a known suspect. Nevertheless, expert analysts can often detect specific characteristics which, in combination with intuitive psychological profiling, can help narrow the field of investigation.

MODERN THEORIES OF CRIMINALITY

British forensic psychologist Paul Britton appropriately described himself as the "jigsaw man." The assessment of criminal psychology is a process of putting together a wide variety of indicative clues in order to construct a picture of the whole personality.

Throughout the 20th century, psychologists continued to explore the root causes of the criminal personality. Nevertheless, although a century has passed since Sigmund Freud's *The Interpretation of Dreams* (1900), a great deal of present-day psychological explanations of the development of violent or sexually deviant personalities remain rooted in Freudian theories.

Freudian theory focuses attention on the importance of how relationships within the family affect how a person develops ways of dealing with others outside the family circle. For example, according to Freud, all male children experience the conflict between loving their father and being jealous of his close relationship with their mother. If the mother is overindulgent or the father is absent, the child cannot resolve his jealousy and eventually develops a hatred of women that can only be expressed by attacks on other women. This certainly appeared to be the motivation of brutal "co-ed killer" Edmund Kemper.

However, Freud's theory does not take into consideration many other factors that lead to a criminal personality, including one that modern-day psychoanalysts have stressed: the point at which such early childhood conflict re-emerges as the characteristic teenage rebellion against both parents. This can result in criminally violent activity, too, if it is not adequately dealt with by the family at an early stage.

The German psychoanalyst Eric Erikson expanded on basic Freudian theory in this respect. As a young man, he met Freud's daughter Anna, who encouraged him to study child psychoanalysis at the Vienna Psychoanalytic Institute.

He moved to the United States in 1933 and taught at Yale and Harvard universities while studying the influence of society and culture on the development of Native American children. His first book,

The German-born psychoanalyst Eric Erikson began by studying the influence of society and culture on the development of young children. In his writings, he identified the occurrence of eight "identity crises" throughout the life of the individual.

published in 1950, was entitled *Childhood and Society*. In it he put forward the concept of the "identity crisis," the conflict that accompanies each step in a growing sense of personal identity.

Erikson defined eight stages in life, during which interpersonal crises need to be resolved:

1: Oral-sensory stage, from birth to 12–18 months. The feeding of the infant results in a loving, trusting relationship with the caregiver, or a sense of mistrust will develop.
2: Muscular-anal stage, from 18 months to 3 years. The child's energies are directed to the attainment of physical skills, such as walking and toilet training. Shame and doubt may develop if this is not handled well.
3: Locomotor stage, from three to six years. The child becomes more independent and assertive, but guilt can develop if aggression is not controlled.
4: Latency stage, from 6 to 12 years. The child is at school and must master new skills, otherwise feelings of inferiority and failure will result.
5: Adolescent stage, from 12 to 18 years. This can be a period of confusion. The teenager has to achieve a sense of identity in sex roles, work, politics, and religion.
6: Young adulthood, from 19 to 40 years. This is a time during which intimate relationships should develop. If not, there will be a growing sense of isolation.
7: Middle adulthood, from 40 to 65 years. Creativity wanes and stagnation threatens. Adults must find some way to relate to the next generation.

8: Maturity, from 65 to death. This is a
time of coming to terms with one's life,
fighting off despair, and developing a
sense of fulfillment.

Each stage of crisis is clearly defined
by a negative element: successively, these
are mistrust; shame and doubt; guilt;
inferiority; confusion; isolation; stagnation;
and despair.

Although Erikson's system of eight
stages has been criticized, it is clear that he
identified the critical points in life, both in
development and in maturity, when a sense
of personal identity is under question and
antisocial behavior or even alienation can
result. In particular, stages 3 to 5 are
those in which a criminal pathological
personality is most likely to emerge if the
identity crises are especially severe.

An example of this is the so-called
"Macdonald triad." Macdonald, an
American psychologist, proposed that the
developing psychopath could be identified
by three progressive childhood traits:
bedwetting, fire raising, and cruelty to
animals.

Most criminal psychologists would not
support Macdonald's generalization,
however, as – although such behavior has
been traced in the history of many serial
killers – this is a striking case of arguing
from the particular to the general.

CONDITIONED BEHAVIOR

In complete contrast with Freudian theory
is the work of the "learning theorists,"
principal among whom was the American
experimental psychologist B. F. Skinner
(1904–90). He believed that a person's
behavior develops as a result of
experiencing the consequences of that
behavior and learning from it. Skinner did
much of his experimental work with
animals, particularly pigeons, which could
be trained to respond in particular ways to
specific stimuli. In effect, this was a
continuation of the experiments into
animal conditioning that were originally
carried out by the Russian psychologist
and Nobel prize winner Ivan Pavlov, early
in the 20th century.

Psychologist and Nobel Prize winner Ivan Pavlov, who pioneered the development of behavioral psychology.

B. F. Skinner theorized that a person's behavior is "conditioned" – changed and developed – by learning from the consequences of that behavior.

We learn to repeat behavior that is "reinforced" and replace that which is made painful. However, Skinner's experimental animals were given a very limited range of stimuli and their rewards – the "reinforcements" – were equally limited. Human beings are able to choose from a wide range of experiences to give them satisfaction. But only seldom after early childhood is behavior shaped purely by reward or punishment. The large number of prison recidivists is evidence of this.

Nevertheless, criminals certainly do learn from their experiences, but this is due less to conditioned responses than to reasoning. The first offender has probably taken advantage of an immediate opportunity, such as a young child grabbing at a toy (or a pigeon suddenly given a reward of grain). However, as he becomes more experienced, he begins to see his criminal actions in a more abstract light and develops a realization of their consequences and implications.

The "Milwaukee Cannibal," Jeffrey Dahmer, committed his first murder without planning and possibly even without premeditation, but he soon developed a technique for finding his victims and drugging them before he killed them. David Canter has called this type of planning "criminal calculus."

Skinner developed the "Skinner box," which was fitted with a lever and a light. An animal placed in the box soon learns that if it presses the lever when the light is switched on it will receive a reward of food. From this and other observations Skinner went on to apply the principle of "operant conditioning" to human beings.

CASE STUDY: JEFFREY DAHMER

Born in Milwaukee in 1960, Jeffrey Dahmer began "experimenting" with dead animals at age 10. He decapitated rodents, bleached chicken bones, and mounted a dog's head on a stake. In June 1978, he was living alone, his parents having separated and left, when he picked up hitchhiker Steven Hicks and took him home for a drink and "some laughs." When Hicks announced that he was leaving, Dahmer crushed his skull with a barbell, strangled him, dismembered his body, and buried the pieces. Thirteen years later he identified Hicks from a photograph, telling the police, "You don't forget your first one."

It seems, however, that this first killing shocked Dahmer into some degree of normality. After a brief spell in college, he signed on for a six-year stint in the U.S. Army but was discharged after only two years due to his heavy drinking. In 1982,

he went to live in a self-contained basement apartment in his grandmother's Milwaukee home, working first for a blood plasma company and later in a chocolate factory. He continued his heavy drinking and his "experiments" with dead animals.

Between September 1987 and March 1988, three young men disappeared without a trace in the Milwaukee area. Dahmer had lured them from gay bars to his grandmother's basement, drugged them, and dismembered them.

By September of that year the stench of his "experimental" activities was too much for his grandmother. She told him to leave,

Skinner carried out many of his conditioning experiments with pigeons. He developed the "Skinner box," in which the birds gradually learned to press a lever when a light switched on, thereby gaining a reward of grain.

Jeffrey Dahmer, the "Milwaukee Cannibal," confessed to murdering 17 boys and men. He preserved many of their skulls and severed heads and claimed to have eaten the flesh of at least one of his victims.

so he moved to a sleazy apartment in the city center.

The day after his move, Dahmer brought a 13-year-old Laotian boy to the apartment, drugged him with sleeping pills, and attempted to seduce him. The boy escaped, the police were informed, and Dahmer was charged with sexual assault. While on bail awaiting sentencing he killed his fifth victim. He kept the head, boiled the flesh from it, and painted the skull. Sentenced to a year in prison to be followed by five years probation, he was

released early in 1990 and soon resumed his killing career.

Between June 1990 and May 1991, he added seven more victims – ranging in age from 18 to 31 – to his total, and preserved their skulls.

Dahmer's 13th victim was nearly his last. He lured a 14-year-old boy to his apartment and persuaded him to pose for photographs in his underwear. Then he drugged him. While the boy was unconscious he decided to slip out to buy more drink. On his return, he found the

half-naked, dazed boy in the street trying to speak to two police officers. Dahmer told them that the boy was his 19-year-old drunk lover, and persuaded them to let him take his "lover" back into the apartment. There he strangled him to death and took more photographs.

Following this close call, Dahmer picked up his next two victims in Chicago, persuading them to accompany him to Milwaukee for photographing. He kept their heads in his freezer and their torsos in a big barrel of acid. Within four days in July 1991, he treated two more victims in the same way. Then, on July 22, a man with handcuffs dangling from his left wrist rushed up to two police officers in a patrol car and told them he had escaped from a "weird dude" who had tried to kill him. He led them back to the apartment – and that was the end of Dahmer's career of slaughter.

The officers found Polaroid photographs of dismembered bodies and skulls and a skeleton hanging in the shower. Opening the freezer, they were horrified to discover four human heads. Following his arrest, Dahmer confessed to the killing of 17 boys and men. The skeleton was that of his eighth victim. He claimed that he had eaten some of the victim's flesh after seasoning it with salt, pepper, and A-1 sauce.

At his trial he was found guilty of 15 counts of murder and condemned to 15 consecutive life sentences, plus 150 years for habitual criminality. He was murdered three years later – in November 1994 – by a fellow prison inmate.

Psychologists have debated the cause of Dahmer's psychopathic nature. Most agree that his home life as a child contributed to

it. His parents argued ferociously, eventually parted, and left him alone. A former classmate described him at 16: "He was lost. He seemed to cry out for help but nobody paid any attention to him at all.... He would come into class with a cup of Scotch whisky. If a 16-year-old drinking in an 8:00 A.M. class isn't calling out for help, I don't know what is." Others have suggested that the fact that he was sexually

Jeffrey Dahmer's apartment contained the putrid remains of at least nine young men.

molested by a neighbor's son at the age of eight contributed to his criminal behavior.

At his trial, Dahmer entered a plea of "guilty but insane," acceptable under Wisconsin state law. Robert Ressler, by now an independent consultant, agreed to appear for the defense, and interviewed Dahmer for two days.

He later wrote: "Dahmer was as candid and as cooperative as any serial murderer whom I have ever confronted, and yet he could not comprehend how he could have committed all of the atrocious deeds that he knew he had done.... There was no way to view this tormented man as having been sane at the time of his crimes."

EYSENCK'S THEORIES

British psychologist Hans Eysenck expanded on Jung's basic distinction between extroverts and introverts during the 1950s. Jung visualized a continuous scale from extroversion to introversion and said that everyone could be placed somewhere along this scale.

Eysenck proposed two main dimensions to each individual personality: one scale running from extroversion to introversion and the other from stable to unstable. This can be visualized as a diamond-shaped diagram. On two sides, the scales run from stable to extroverted and from extroverted to unstable. On the other two sides they

Far left: At his trial, Dahmer pled guilty but insane. In February 1992, the jury found him sane and responsible for his actions, and the court sentenced him to 15 consecutive life sentences.

Below: British psychologist Hans Eysenck, who regarded the criminal personality as being a combination of hereditary and environmental factors.

run from stable to introverted and from introverted to unstable.

The four sides of this diamond are named with personality types that were first distinguished some 500 years ago:

> **Hans Eysenck concluded that crime could be a rational choice, given the possibility of maximizing pleasure and minimizing pain.**

- stable to extroverted: sanguine
- extroverted to unstable: choleric
- stable to introverted: phlegmatic
- introverted to unstable: melancholic

Most people fall somewhere in the middle of one range. The sanguine type reveals qualities of leadership, is lively and easygoing, talkative, and sociable. The choleric type progresses toward instability through impulsiveness, excitability, aggression, and restlessness, and can be

very touchy. The phlegmatic type is calm, even-tempered, and thoughtful, but as the tendency to introversion increases he or she becomes overly careful and passive. The melancholic type progresses through unsociability, pessimism, rigidly held opinions, anxiety, and moodiness. Consequently, unstable personalities will be of either the advanced choleric or melancholic types.

Eysenck has been severely criticized, at least partly, because he based his theories on the assumption – which he claimed to have established experimentally – that individuals differ genetically in their learning abilities, particularly their ability to respond to environmental conditioning. He also assumed that crime could be a natural, rational choice for certain individuals, given the possibility of maximizing pleasure and minimizing pain. He therefore believed that the criminal personality was based upon a combination of hereditary and environmental factors.

Eysenck was later forced to introduce a third dimension of personality, which he named psychoticism: the individual tends to be solitary, cruel, and sensation seeking. Exactly how this dimension fits in with those of extroversion–introversion and stable–unstable is not clear, but Eysenck associated it with criminality. Some support for his theories has been found in a

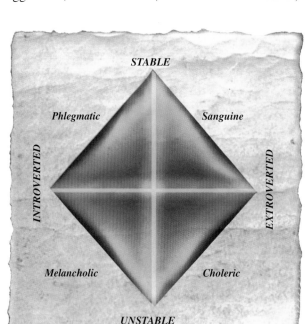

A diagram representing Eysenck's scale of personality types. Most people can be characterized as occupying a position on one of the four sides of the diamond.

STABLE

Phlegmatic   Sanguine

INTROVERTED   EXTROVERTED

Melancholic   Choleric

UNSTABLE

BIOLOGICAL FACTORS

Hormones are "messenger" chemicals secreted into the bloodstream by endocrine glands or nerve cells. The messages they carry regulate the growth or functioning of specific organs or tissues – and in the process they may influence human behavior. In the last half-century, researchers have focused on the possibility that an imbalance in the level of sex hormones in the body could be connected with criminality.

Research has suggested that the male sex hormone testosterone is directly related to aggressive, antisocial crimes such as rape and murder. Testosterone levels peak during puberty and the early 20s, and this is the time associated with the highest crime rates.

It is also known that the limbic system of the brain, which controls emotions such as rage, love-hate, jealousy, and religious fervor, can be affected by testosterone. However, although the most violent rapists have been found to have high testosterone secretion, no direct relationship has been established between criminal behavior and sex hormone levels.

Another important hormone is epinephrine. It raises the rate of blood flow, improves the power of the muscles, and increases the rate and depth of breathing – preparing the body for "fright, flight, or fight," a state known as cortical arousal. This is what can be measured by the "lie detector," because most people show a fear of being revealed as liars.

Scientists have discovered that those with low cortical arousal are often habitually aggressive. And when threatened with pain, criminals show less stress and cortical arousal than other people.

One group of researchers reported that some criminals, particularly violent ones, needed stronger stimuli to arouse them than non-criminals, and they recovered more slowly to normal levels. They linked rapid recovery to the ability to learn from unpleasant stimuli, such as punishment, and postulated that criminals are less able to learn acceptable behavior from either negative or positive stimuli.

Researchers have also looked at the effects of nutritional deficiency or imbalance on behavior. For example, a recent study reported that a group of prison inmates who had been given a specially designed diet with enhanced quantities of vitamins showed a marked improvement in their general behavior.

Blood sugar levels proved remarkably important in a case of murder in 1989. A man stabbed his wife, then attempted to commit suicide by slitting his wrists. He was subsequently found to be suffering from amnesia. For two months preceding the murder he had been on a very severe diet, denied sugar, bread, potatoes, and fried food, but on the morning he killed his wife he had drunk two glasses of whisky.

At his trial, medical experts testified that he was suffering from hypoglycemia – an abnormally low blood sugar level. They said that this was sufficient to impair normal brain function and substantially reduce his sense of responsibility, "perhaps even to render him an automaton." The jury found him not guilty.

recent study that included both criminals and non-criminals. Unstable extroverts and unstable psychotic extroverts were found only among the criminals. Stable introverts were found only among the non-criminals. Unstable introverts and stable extroverts occurred in both groups.

THE CRIMINAL PERSONALITY

For many years, psychologists have searched for methods to assess and define the many aspects of human personality and establish standards of normality. At the same time, groups of researchers have struggled to determine if generalizations can be made about the "normal" criminal personality and how far this differs significantly from the norms of the population as a whole. Researchers have compared personality tests of known criminals with "non-criminals," who are assumed to represent the common norm.

There are over 30 different types of personality tests. They include IQ assessments, free association tests, the Rohrschach test (in which the subject has to interpret the inherent symbolism of a symmetrical inkblot), and – what has been claimed to be the most reliable in assessing the criminal personality – the Minnesota Multiphasic Personality Inventory (MMPI).

TESTING PERSONALITY

The MMPI was developed in the 1950s. It is a questionnaire of 550 statements that subjects have to decide as being true or false when applied to themselves – and there are a number of cross-checks built into the questionnaire that are designed to detect untruthful answers.

The test is divided into 10 scales and the subject is given a score on each scale. There is no overall score. Each of the 10 scores is plotted on a graph from which the subject's personality is assessed (see box below).

The Pd scale (4) was specifically designed to distinguish delinquents from other groups, and a study in 1967 found that it most often differentiated criminals from non-criminals. However, those who had dropped out of school, whether delinquent or not, were found to have a higher Pd score than others, as were ambitious young, upwardly mobile professionals, aggressive businessmen, professional actors – and those who had accidentally shot someone in a hunting accident. A more recent study (in 1993) suggested that a combination of the scores on scales 4 (Pd), 8 (Sc), and 9 (Ma) could

MMPI SCALE RESULTS

Scale no.	Scale name	What the scale discloses as the score rises
1	Hypochondria (Hs)	Tired; inactive; lethargic; feels physically ill
2	Depression (D)	Serious; low in morale; unhappy; self-dissatisfied
3	Hysteria (H)	Idealistic; naive; articulate; social; ill under stress; psychosomatic symptoms
4	Psychopathic deviation (Pd)	Rebellious; cynical; socially aggressive; selfish
5	Interest pattern of opposite sex (Mf)	High: sensitive. Low: own sex interest pattern. High score in males: gentlemanly; scholarly; feminine. High score in females: rough; ambitious
6	Paranoia (Pa)	Perfectionist; stubborn; hard to know. Low: socially acceptable
7	Psychasthenia (Pt)	Dependent; desires to please; feels inferior; indecisive; anxious
8	Schizophrenia (Sc)	Negative; difficult; pathetic; lacking social graces
9	Hypomania (Ma)	Expansive; optimistic; decisive; not bound by custom
10	Social introversion (Si)	Unassertive; self-conscious; shy. Low: socially acceptable

I-7: full awareness of the integrating process between self and others.

A person who reaches full social maturity is one who has successfully passed through the earlier identity crises postulated by Erikson.

Studies in the 1970s found that convicted offenders were all, to a degree, socially immature, falling on levels 2, 3, or 4.

The I–2 level represents a person concerned solely with his or her own needs, unable to understand or predict the reactions of others. Those at this level have been divided into "asocial, aggressive" (actively demanding and aggressive when frustrated) and "asocial, passive" (complaining and withdrawn when frustrated).

At level I–3, people show some awareness of the effects of their behavior on others; consider the environment as something that can be manipulated by the exercise of power; and rely on strict "either is or is not" rules. They are divided into "passive conformists," who will follow the lead of whoever has power at the moment; "cultural conformists," who model their behavior on a specific delinquent group; and "asocial manipulators," who attempt to undermine authority for their own ends.

I-4 people are concerned with status and respect. They imitate the roles of others, possibly identifying with hero figures, and they set up rigid standards for themselves that can produce feelings of inadequacy.

A Rohrschach inkblot has no inherent representational content and is generated by folding a sheet of paper over a drop of ink. Those taking the test are asked what they can "see" in the resultant pattern, and their answers can give the investigating psychologist an insight into their personalities.

be more successful in predicting criminality, but the practicality of such tests in criminology remains doubtful.

THE "I-LEVEL" TEST

Another class of personality test is the "interpersonal maturity" or "I-level" test. Individuals are tested for their social skills and placed on one of seven levels of developing maturity in ascending order. Briefly stated, the seven I-levels are:

I-1: basic awareness of self, as distinct from non-self
I-2: differentiation between persons and objects
I-3: distinction of simple social rules
I-4: awareness of the expectations of others
I-5: empathy with others and under-standing of differentiation of roles
I-6: differentiation between self and social roles

THE KILLER CLOWN

John Wayne Gacy, who liked to dress as a clown for charity events in Des Plaines, Illinois, was found guilty of murdering and dismembering 33 boys and young men in 1980.

He appealed and, while in prison, planned his own funeral service, naming the hymns to be sung and the type of coffin he wanted to be used.

In June 1988, the FBI held its first

International Homicide Symposium at Quantico, and Robert Ressler arranged a closed-circuit pair of satellite interviews with Edmund Kemper and Gacy. Kemper was candid about his crimes, but Gacy spent all 90 minutes of his interview trying to persuade the audience of law-enforcement officials that he was innocent, and that they should support his appeal.

Far right: Seventeen-year-old William Heirens killed two women in their apartments in 1945 and subsequently kidnapped and dismembered a six-year-old girl. After his second murder, he left a message in lipstick on a wall: "For heaVems Sake catch me Before I Kill More I camnot control myselF." He claimed the murders were committed by a different element of his personality, which he called "George Murman."

Those characterized as "neurotic" may act out their guilt reactions to suppress conscious awareness of this inadequacy or become emotionally disturbed by anxiety. Others act out an immediate response to family or personal crises. A fourth group lives by their own delinquent beliefs.

The reported results of both the MMPI and I-level tests suggest that there is a distinguishable connection between criminality and assertiveness, hostility, resentment of authority, and psychopathic behavior. However, there is an inherent drawback to the research that has been carried out using either of these tests. Every study compared convicted criminals with non-criminals. The "non-criminal" group included those not convicted of any crime. But such a group might well contain criminals who had successfully avoided arrest and conviction (a reservation also pointed out by Brent Turvey).

In addition, some of the personality

characteristics found in convicted criminals could well derive not from something inherent in their natures or developed by childhood conditioning but from their treatment by the justice system, which could certainly result in resentment of authority.

CHILDHOOD TRAUMA

By far the most disturbing of recent trends in crime has been the increased incidence of serial murder, particularly in the United States. A number of American researchers insist that the cause can be found, as most of the psychological theories outlined above suggest, in severe childhood neglect or abuse. They claim that the newly formed fetus in the womb can be affected by the mother's alcoholism or drug use, or even by the psychological strain of an unwanted or unhappy pregnancy.

One effect of early childhood trauma is "dissociative identity disorder" (DID),

KILLER STATISTICS

Between 76 and 85 percent of cases of serial murder have occurred in the United States, much of the rest having taken place in Europe. These figures are open to question because relatively few cases have been reported from Russia, the Middle East, and Asia.

The highest figure for serial murder in the United States comes from California, with Texas, New York, Illinois, and Florida close behind. A the other end of the scale, Hawaii, Montana, North Dakota, Delaware, and Vermont are credited with just one case apiece, and none has been reported for Maine.

Of the European total, England has produced 28 percent, Germany 27 percent, and France 13 percent.

- More than 90 percent of the world's serial killers are male.
- 84 percent of serial killers are white, 16 percent black.
- 26 percent begin killing in their teens, 44 percent in their 20s, and 24 percent in their 30s.
- 86 percent of all killers are heterosexual.
- 89 percent of victims are white and 65 percent are female.

Richard Ramirez, the "Night Stalker" of Los Angeles, caused a reign of terror from June 1984 through August 1985. At his trial, he boasted: "I am beyond good and evil.... Lucifer dwells in us all."

which is described as a "creative survival technique" that a child employs to escape from physical, sexual, or emotional abuse. "During dissociation," one psychologist has written, "one is not able to associate certain information as one normally could, thus allowing a temporary mental escape from the fear and pain of experience.

"This process can, at times, result in a memory gap concerning the trauma, which may affect the person's sense of personal history and identity, and may even result in fragmenting one's self." This effect, which clearly echoes Erikson's theory of identity crisis, can continue into adult life. It is illustrated in Paul Britton's early advice on the interrogation of Paul Bostock.

Modern psychologists agree that the potential for criminal violence is engendered by childhood trauma within the family. Eventually the child learns that a violent response to abuse earns him respect and fear, and in this way he develops the belief that violence is his best way of dealing with people.

The argument is persuasive, although it is difficult to accept that the level of childhood abuse in the United States is so much higher than in other countries that American serial murder makes up more than three-quarters of the world total (see box, page 214). Indeed, there is fragmentary evidence to suggest otherwise. For example, the early 20th-century pathological criminal Carl Panzram, writing from an American prison cell in 1930, claimed to have murdered 21 people and committed thousands of burglaries, robberies, larcenies, and arsons.

Yet he said: "All of my family are as average as human beings are. They are honest and hard-working people. All except me. I have been a human animal ever since I was born. When I was very young, 5 or 6 years of age, I was a thief, a liar, and a mean despicable one at that. The older I got, the meaner I got."

Nevertheless, in the absence of other convincing theories, the figures must be left to speak for themselves.

Dr. Lonnie Athens, an American criminologist, carried out an extensive study of violent criminals in prison and claimed to identify a pattern of social development common to all – a four-stage process that usually occurred within the family that he has named "violentization." He based much of this upon his own childhood experiences. The four stages are:

1: Brutalization. The young child is forced to submit to an aggressive authority figure by violence or the threat of violence. He witnesses the violent subjugation of his intimates and learns from the authority figure that violence can be used to settle disputes.

2: Belligerency. The child determines to avoid further violent subjugation by himself resorting to violence.

3: Violent performance. His violent response to provocation succeeds and he earns respect and fear from others.

4: Virulency. Flushed with success, he decides that serious violence is the best way of dealing with people and he begins to associate with others who believe the same.

THE POSSIBILITY OF TREATMENT

Given the many lines of research described above, the question naturally arises: would it be possible to treat convicted offenders – particularly killers and rapists – in some way that would ensure that they could be released into the community and would not offend again? Or, equally importantly, could the potential for criminality be detected at a sufficiently early age so steps could be taken to halt its development?

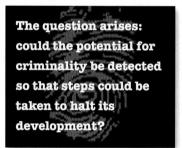

The question arises: could the potential for criminality be detected so that steps could be taken to halt its development?

Unfortunately, current methods of psychological treatment of offenders in institutions do not appear to be effective. There have been far too many cases in which violent offenders – Edmund Kemper and Henry Lee Lucas are prime examples – have been declared "cured," only to resume and often accelerate the execution of their crimes.

The assumption that high testosterone levels were the prime causal factor in rape and other sexual offenses led many European countries during the 1920s and

1930s to legalize castration for sex offenders. Although thousands of such operations were carried out, it was never established that they in any way reduced or curtailed sexual aggression.

In the last 30 years, specific drugs that reduce some of the effects of testosterone were offered to offenders as an alternative to castration. In some cases, these were found to be effective, but they have to be administered in high doses and do not appear to offer a definite cure.

Sociologists wish that the early development of psychopathic tendencies in children could be detected and reported, and that they could obtain official authorization to apply behavioral treatment. However, this would require a wide-ranging and extremely expensive program for which neither funding nor the acceptance of justification for such a degree of social interference currently exists. It therefore seems that, at present, there is no way of detecting and halting the development of criminal personality, until it is revealed in the commission of a crime – at which point the profiler comes into his own.

Since December 1985, a Special Unit at Parkhurst Prison in Britain has treated violent criminals. The consultant psychiatrist in the unit described his work as "disentangling the long-term effects of childhood abuse." He discouraged the use of tranquilizing drugs, which medical research showed could induce violent behavior.

NEGOTIATION AND INTERROGATION

Even after an offender has been pinpointed, the criminal profiler has a great deal to contribute to the conclusion of a case. Crisis negotiation is of critical importance. The FBI has opened a new operations center (left) at their headquarters in Washington, which is capable of dealing with five crises at once.

No criminal case is complete until the offender has been apprehended, tried, found guilty, and sentenced. So far, this book has been concerned with what can be done to lead the investigators of a crime toward the identification of an unknown offender (UNSUB) by analyzing his or her psychological characteristics. There are, however, circumstances in which one or more offenders have been contacted or located – even if not yet identified – in kidnapping, hostage-taking, or barricaded suicide cases. In all these situations, too, an understanding of criminal psychology is essential.

Very soon after FBI agents Howard Teten and Pat Mullany began to teach the elements of criminal psychology analysis at Quantico, they realized that very similar concepts could be applied to these siege-type situations. Howard Teten wrote: "In 1974, Pat Mullany and I developed a hostage negotiation program, using the principles underlying the profiling process." During 1974 and 1975, Mullany negotiated several hostage-taking circumstances, making use of these newly developed techniques. Later, his work was modified and expanded by FBI agents Con Hassel and Thomas Strentz.

An experienced understanding of criminal psychology can also be valuable in the interrogation of apprehended suspects.

Some of the 21 hostages abducted from Malaysia by Philippine Abu Sayyaf rebels in May 2000 being held at the rebels' jungle camp. At one time the rebels held more than 70 hostages, including two Americans: some were released after a few months, but others were found dead. U.S. advisers said that Philippine troops failed to effectively communicate in negotiations with the rebels.

Accredited profilers, as well as consultant psychologists or psychiatrists, can play a critical role in advising the police (or, in certain cases, representatives of insurance companies) on the best way to pursue both negotiation and interrogation.

In fact, the basic strategy is very much the same in all these situations: the natural response of direct confrontation must be restrained and be replaced by attempts to persuade the subject – whether he or she is an arrested suspect; a person threatening to commit suicide, or armed and barricaded in a building; a kidnapper; or a terrorist – that the negotiator has his or her interests at heart. A relationship of trust and mutual

sympathy must be developed, based upon an understanding of criminal psychology: this will not only make it easier to explore ways of resolving the situation, but will also, by introducing an element of calm discussion, reduce the likelihood that the subject will suddenly express frustration in an act of violence.

NEGOTIATION

The situations in which the police, or other security organizations, need to deal with people threatening violence to themselves, or others, cover a wide range. Although these situations may appear very different in nature, they are – as FBI Special Agents

DOG DAY AFTERNOON

At closing time at a small bank in Brooklyn, New York, Sonny (Al Pacino) and his weak-willed accomplice Sal (John Cazale) uncover their weapons and announce a hold-up. But there is little available cash, the alarm has been given, and the two find themselves holding the bank staff hostage – while outside crowds gather, police sharpshooters take up positions around the building, and Lt. Moretti (Charles Durning) is faced with keeping the situation as low-key as possible. As the long, hot afternoon turns to evening, Moretti – while being supervised by FBI agent Sheldon (James Broderick) – negotiates with Sonny by telephone; at the same time, the bank employees begin to establish some degree of empathy with the robbers, expressing understanding of their problems and even sympathizing. Pizzas are delivered to the bank, and eventually the police agree that the gunmen, with their hostages, will be taken by limo to the airport, where a plane waits to take them out of the country. It is only at the airport that the FBI and police intervene: Sal is suddenly shot dead at pointblank range, and Sonny – pressed up against the side of the limo – realizes that all his plans have come to nothing.

Al Pacino in the role of Sonny, in the movie made by Sidney Lumet. Based on a true event in New York, *Dog Day Afternoon* became an important element in the training of BSU negotiators.

Previous page: Officers from a police SWAT team take up position during a training exercise in Mt. Zion, Georgia.

Dwayne Fuselier and Gary Noesner pointed out in a 1990 paper – all "criminal acts," and the approach to negotiation in every case should remain fundamentally the same. What was originally called "hostage negotiation" is now covered by the broader term of "crisis management."

The natural reaction of the authorities to any of these situations is confrontation: in cases of threatened violence, among early arrivals at the scene are likely to be armed police, followed, in more extreme developments, by specialist sharpshooters, or even SWAT (Special Weapons and

Tactics) teams. This "knee-jerk" reaction must be rigorously restrained: any overt physical action will inevitably result in a response that places the lives – of the offenders themselves, any hostages, and indeed the police officers – at risk.

The technique of how to deal with a person on a high ledge, threatening to jump, is the same in every case: no immediate attempt to seize them, but a sympathetic discussion of their problems and causes, and the gradual establishment of mutual trust; the suggestion that there are other ways to deal with these problems; the careful, surreptitious containment of the scene; and then – and only then, if any other resolution seems impossible – the sudden firm grasp of the ankle or arm. Even a siege situation, in which a considerable number of hostages is being forcibly held by a group of armed terrorists, should be approached in substantially the same way – the final stage, in this case, being a very sudden, unheralded, physical assault.

Specific FBI instruction in hostage negotiation techniques began in the BSU in 1975 under the supervision of Robert Ressler. Some 50 agents took the first course, with Pat Mullany as senior instructor, and Captain Frank Bolz and Detective Harvey Schlossberg of the New York Police Department (NYPD) providing advice based on their practical experience. John Douglas wrote: "We studied some of the significant phenomena that had arisen out of hostage situations, such as the Stockholm syndrome." One of the NYPD cases had been made into a film by Sidney Lumet – *Dog Day Afternoon*, starring Al Pacino. It was based on a real-life hostage incident that took place in New York City on August 22, 1972, in which a man attempted to rob a bank to get money to pay for a sex-change operation for his male lover. The movie was studied by the FBI as part of the course.

One of Pat Mullany's early successes came when Cory Moore, a man diagnosed as a paranoid schizophrenic, took a police

Throughout the 1980s, the FBI carried out a program of realistic training exercises, in which experienced BSU and VICAP operatives played the part of hostage takers. In this photograph, Robert Ressler (right), disguised as a terrorist, is seen making demands in a typical scenario.

In 1990, a gunman entered the Mormon Church Genealogical Library in Salt Lake City, Utah, and killed two people. Following a firefight, the man was found dead; police said that they did not know whether he had been shot or had died from self-inflicted wounds.

captain and his 17-year-old secretary hostage inside the police station at Warrensville Heights, Ohio. Among the gunman's demands were that all white people should immediately leave planet Earth, and he said he wanted to discuss this with President Jimmy Carter. Robert Ressler, who was also present, took a call from the White House press secretary announcing that the president was willing to talk to "the terrorist."

But as John Douglas wrote in

Mindhunter: "Once you put the hostage taker in direct contact with someone he perceives as a decision maker, everyone is backed against the wall, and if you don't give in to his demands, you risk having things head south in a hurry. The longer you keep them talking, the better." Ressler lied, telling the White House that the gunman could not be reached presently by telephone. Mullany was able to resolve the situation without bloodshed – or presidential intervention. He offered

Moore a press conference at which he could explain his views, and the hostages were released.

Crisis management teams are now an important part of many law enforcement organizations throughout the world. They typically include two or more equipment technicians, experienced in telephone systems and surveillance equipment – the telephone is now the preferred method of communication with the subject – and a number of specially-trained negotiators, who regularly update their skills in refresher courses. Backing them up, if necessary, is an armed "tactical unit" – but negotiators stress that their presence must be concealed, and that force should be used only when all else fails.

ARMED RESPONSE – THE LAST RESORT

The first SWAT teams in the United States were drawn from ex-soldiers returning from Vietnam in the late 1960s and early 1970s who joined police forces. With training in weaponry and assault tactics, these teams were paramilitary forces. This was an aspect that not even the FBI had previously considered. Unfortunately, the "gung-ho" attitude of many of these teams

Police carry out a hostage rescued during a siege at the Japanese ambassador's residence in Lima, Peru, in April 1997. At least one hostage was killed, and all the hostage takers died when police stormed the building.

clashed with their standing as "peace officers." They employed snipers and even grenade launchers in their efforts to storm criminal hideouts and rescue hostages. Many criminals were killed, but so were many police officers, and hostages were also frequently injured during the firefight. The NYPD was the first to face up to the problem by instituting a policy of negotiation. The FBI followed their example by commissioning a study of negotiation techniques during the 1970s.

In *Mindhunter*, John Douglas highlights an example of what they set out to avoid. Jacob Cohen, a man wanted for killing a police officer in Chicago, was traced to an apartment building in Milwaukee where he shot and wounded an FBI agent who tried to approach him. A newly formed and relatively inexperienced FBI SWAT team surrounded the place, but Cohen ran through them – taking two bullets in the buttocks. He grabbed a young boy and burst into a house occupied by an adult and a child. Cohen now had three hostages.

Tempers ran high. The Milwaukee police and the FBI blamed each other for allowing the situation to deteriorate. The SWAT team was angry for letting Cohen get through. Cohen was in a rage and in considerable pain. To complicate matters, the Chicago police announced that they too were coming to get their man.

Then further mistakes were made. First, the senior FBI agent present at the scene loudly harangued Cohen with a bullhorn. He offered himself as a hostage in exchange for the boy and began to drive his car slowly toward the house. Meanwhile, FBI Agent Joe DelCampo climbed onto the roof. Cohen came out of the house with his arm wrapped around the boy's head. Suddenly the boy slipped and fell. At the same moment, DelCampo fired and Cohen also fell – but nobody knew whether he or the boy had been hit.

Immediately everybody opened fire. The FBI car was riddled with bullets. The agent inside and a police detective were injured and Cohen was hit with more than 30 shots. Miraculously the boy survived, but he was injured when the out-of-control FBI car rolled forward onto him. "But," as Douglas wrote, "that wasn't the end of it. Fistfights nearly break out and the police almost beat up DelCampo for taking their shot." This was clearly no way in which to deal with a hostage situation, and it provided strong justification for the FBI's ongoing study of crisis management.

> In an attempt to legally define "terrorism," FBI experts have suggested the less emotive term "planned political/ religious hostage taking."

DEALING WITH TERRORISTS

The term "terrorist" was overused in the 20th century. As FBI agents Fuselier and Noesner pointed out in their 1990 paper: "Both the general population and the law enforcement community have come to accept the terrorist stereotype as accurately depicting personality traits, dedication, sophistication, commitment, and modus operandi.... The words 'terrorist' and 'terrorism' have been used by the media to such an extent that they are virtually useless as valid descriptive terms."

The FBI defines terrorism as "the unlawful use of force or violence against persons or property to intimidate or coerce a government, civilian population, or any segment thereof, in furtherance of political or social goals." Fuselier and Noesner suggest that a less emotive – and more descriptive – term might be "planned political/religious hostage taking." They believe that "terrorists" should not be differentiated from the broad range of criminal and psychologically disturbed personalities more frequently encountered in crisis management.

One of the greatest problems facing law enforcement officials who have to attempt negotiation with hostage-taking terrorists is that there is no detailed understanding of terrorist motivation and no justification for the popular view that terrorists are just "crazy psychopaths." Indeed, David Long, former assistant director of the U.S. State Department's Office of Counter Terrorism, concluded that terrorists do not even share a common personality type. "No comparative work on terrorist psychology," he wrote, "has ever succeeded in revealing a particular psychological type or uniform terrorist mindset."

Nevertheless, Long agrees with others that rank-and-file members of terrorist organizations tend to have a low level of

Armed German police take up positions outside the Olympic village dormitory in Munich where 11 Israeli athletes had been seized by Palestinian terrorists in September 1972. The disastrous finale to the operation, in which all the hostages died, led to a radical reassessment of the training of SWAT groups in Germany.

Algerian fundamentalists hijacked an Air France plane at Marignane airport, Marseille, in December 1994. Here the passengers are being escorted to safety moments after members of a French elite anti-terrorist unit had stormed the plane, killing all four hijackers.

self-esteem and are therefore drawn to charismatic leaders. This matter of self-esteem is important. As John W. Crayton argues in *Perspectives on Terrorism* (1983), humiliating terrorists in siege situations by withholding food, for example, is counterproductive because "the very basis for their activity comes from their sense of low self-esteem and humiliation."

Psychologist Rona Fields, who has spent 30 years researching the mentality of terrorists, says: "Their definition of right and wrong is very black-and-white, and is directed by an authoritative director. There's a total limitation of the capacity to

think for themselves." She found that they were more likely to be recruited in early adolescence, driven by blind desire for revenge – "an eye for an eye" – over injustice or violence suffered by those near to them.

Stephen J. Morgan writes in *The Mind of a Terrorist Fundamentalist* (2001): "Children like this... are love cripples, emotionally deformed and lacking in anything but the most artificial shell of a normal personality…. Inside themselves they carry emotional bombs, which tick away for years, until the time comes for them to detonate themselves and kill everyone around them."

HOSTAGE NEGOTIATION TRAINING

From 1977 and throughout the 1980s, full-scale, week-long simulations of terrorist attacks and hostage negotiation were held at isolated sites in the United States in an attempt to refine the principles first developed by the BSU and improve understanding of the psychological pressures involved in siege situations. The people who took part included representatives from the FBI, the CIA and the U.S. Army's Delta Force, as well as Special Air Service (SAS) men from Britain and specialists from other foreign organizations. On a number of these exercises, Robert Ressler played the part of chief terrorist: "We would hijack a busload of volunteers who played important people – scientists or visiting dignitaries, for example – and take them to an isolated farm or ski lodge, where we held them hostage. Real guns, grenades, dynamite and other weapons were used, and when I demanded an airliner to fly us out of the country, one was commandeered and delivered to the nearest airstrip…. These simulations were so realistic that some

THE FEDERAL RESEARCH DIVISION REPORT

Following the Al-Qaeda attacks on the U.S. embassies in Nairobi and Dar-es-Salaam in 1998, the National Intelligence Council commissioned a report on "The Sociology and Psychology of Terrorism" from the Federal Research Division in June 1999. Completed three months later, the report exemplifies the problems faced by analysts in defining the threat of international terrorism.

The preamble promised "to develop psychological and sociological profiles of foreign terrorist individuals and selected groups to use as case studies in assessing trends, motivations, likely behavior, and actions that might deter such behavior, as well as reveal vulnerabilities that would aid in combating terrorist groups and individuals." However, the authors admitted to being "limited by time restraints and data availability," and the data was restricted to a survey of psychological research carried out on terrorism in general and a series of brief descriptions of the careers of a few known terrorists.

The horrific results of the Al-Qaeda truck-bomb suicide attack on the U.S. embassy in Nairobi, Kenya, on August 7, 1998.

Members of a police tactical team involved in the search for the "Beltway sniper" in October 2002. Initial profiles of the alleged offender were found to be far off the mark when two perpetrators, John Allen Muhammad and John Lee Malvo, were eventually arrested.

hostages became subject to the 'Stockholm syndrome'...."

Those who played the part of negotiators during these exercises sometimes protested that Ressler's experience made him too powerful an adversary, because he was familiar with all the psychological negotiating techniques that he had taught them.

THE STOCKHOLM SYNDROME

Among the problems that negotiators may encounter in a hostage situation is the way in which the hostages can develop sympathy with their captors, to such an extent that it can interfere with the peaceful resolution of the situation. This has been named the "Stockholm syndrome" by sociologists, following the unexpected reaction of four Swedish bank employees held captive for nearly six days in August 1973. Detailed psychological study of this and similar cases has resulted in the definition of a characteristic set of symptoms that behavioral experts have to take into consideration in the training they give to negotiators.

Early in the siege, the captives begin to identify with their captors. At first, this may be no more than a defensive response, based on the feeling that the captor is less likely to harm the captive if he is cooperative and apparently supportive. This engenders an almost childlike desire on the part of the captive to secure the favor of the captor.

As the siege continues, the captive begins to realize that any action taken by his would-be rescuers could result in harm to himself: if a firefight develops, he may be wounded or killed, whether by the attacking authorities or by the infuriated captor. He becomes increasingly opposed to any violent resolution of the situation.

Identification with the captor grows as time passes: he reveals himself as a similar human being, with his own hopes and aspirations. The captive becomes gradually familiar with the captor's point of view, his grievances, and his ideological or political beliefs. In due course, the captive may well become persuaded that the captor's cause is just, placing the blame for his situation on the authorities and the rescue team.

Increasingly, the captive tries to distance himself emotionally from the situation by denial. He may try to persuade himself that "it is all a dream," losing himself in long periods of sleep, or in delusions that he is about to be rescued by some "magical" means. Sometimes he finds opportunities to deny his situation by indulging in pointless – but time-consuming – activities related to his occupation before being taken hostage.

Psychologists have concluded that the Stockholm syndrome is present in cases other than hostage situations, such as the persistent emotional attachment that is seen in cases of domestic violence. When someone threatens your life, deliberates, and does not kill you, wrote one psychologist, the relief at removal of the threat generates intense feelings, both of gratitude and of fear. These combine to make the victim reluctant to reveal any negative feelings toward the threatener. "The victim's need to survive is stronger than his impulse to hate the person who has created his dilemma," wrote FBI profiler and hostage negotiator Thomas Strentz in 1980. Researchers have summarized the stages in which the Stockholm syndrome can develop:

A person threatens to kill another and is perceived as having the ability to do so.

The other cannot escape, so his or her life depends upon the threatening person.

The threatened person is isolated from outsiders, so that the only other perspective

The 11 x 47 foot vault in which four employees of the Sveriges Kreditbank, Stockholm, were imprisoned by two escaped convicts on August 23, 1973; and one of the hostage takers, arrested four days later by Swedish police wearing gasmasks. Investigators were surprised that the hostages actively resisted their rescue by the police, refused to testify against their captors, and even raised money for their legal defense. It was reported that one of the female hostages later became engaged to one of the imprisoned men. Their reaction gave rise to the definition of the "Stockholm syndrome."

PATRICIA HEARST

One of the strangest examples of the Stockholm syndrome is the case of Patricia Hearst, the 19-year-old granddaughter of the newspaper mogul William Hearst, and her involvement with the tiny American terrorist group calling itself the Symbionese Liberation Army (SLA). The SLA was the brainchild of Donald DeFreeze, who escaped from prison in March 1973 and was joined soon after by a handful of "revolutionaries," who began to steal guns and rent hideouts.

In November 1973, the group's first action was the assassination of Dr. Marcus Foster, the Oakland schools superintendent whom they accused of "fascism." Not long after this they kidnapped Patricia Hearst. The kidnappers demanded that her parents provide $70-worth of food to each poor person in California, which by their calculation made a total of $400 million. The Hearsts offered $2 million in advance, and a further $4 million if their daughter was released unharmed. Some food was distributed in a program named "People in Need," but the SLA responded with a tape-recorded message from Patricia Hearst – along with a photo of her carrying a submachine gun in front of the SLA's symbol, a seven-headed cobra. On the tape, she announced that she had taken the name "Tania," and that she had decided to remain with the SLA "and fight."

On April 15, 1974, five members of the SLA, including "Tania," were seen on video robbing a bank in San Francisco. On May 16, she covered the escape of two shoplifting comrades in Los Angeles by firing shots into the air. The next day investigators discovered an SLA hideout in Los Angeles: 400 police officers and FBI agents shot it out with the six occupants before the house burned down. "Tania" was not among the SLA members captured. The SLA was effectively finished but its remaining members were not captured until September 1975.

At her trial, defense attorney F. Lee Bailey argued – unsuccessfully – that Patricia Hearst had been brainwashed during her captivity. She was found guilty and sentenced to seven years in jail in March 1976. President Carter commuted the sentence in February 1979, and on January 20, 2001, President Clinton granted Patricia Hearst a full pardon.

available is that of the threatening person.

The threatening person is perceived as showing some degree of kindness to the one being threatened.

PSYCHOLOGICAL ASPECTS OF INTERROGATION

Police interrogators often use the "bad cop/good cop" approach. The "bad cop" submits the suspect to an aggressive grilling, while the "good cop" remains quietly in the background. At a previously agreed point in the interrogation the "bad cop" finds some excuse to leave the interview room and the "good cop" takes over. He expresses concern at the way in which the suspect is being treated by his colleague, offers him a cigarette, a glass of water or a soda, or even a sandwich. He remarks sympathetically how much easier it would be for the suspect to "come clean." Sometimes this is sufficient to produce an immediate confession. More often, when the "bad cop" returns and resumes his attack, the suspect may say: "I won't talk to you – but I will to him." Eventually, sometimes after a succession of exhausting sessions, the truth emerges.

A former variation on this approach, now infrequently used, was the employment of a stool pigeon. Another prisoner willing to help the police – or even a police officer pretending to be a prisoner – was locked into the same holding cell as the suspect, whom he encouraged to talk about the crime for which he or she had been arrested.

In the past police sometimes employed what was known as the "third degree," named after the detailed questioning directed at a candidate for the degree of Master Mason (the third) in Freemasonry.

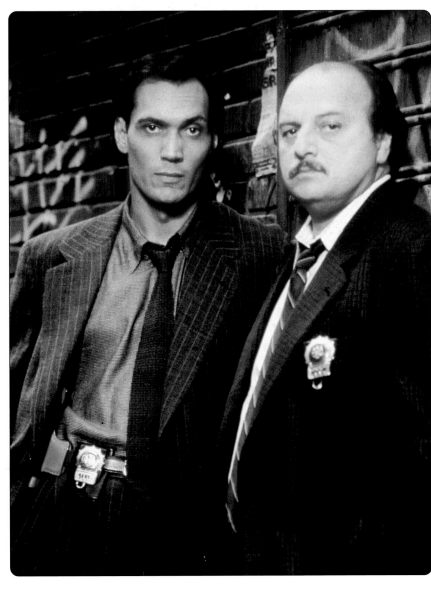

Police detectives Simone and Sipowicz (played by Jimmy Smits and Dennis Franz) on the popular TV show NYPD Blue *regularly employ the bad cop/good cop routine when interrogating suspected criminals.*

LIE DETECTORS

At one time, great things were expected of the polygraph "lie detector." Its development was based on the belief that a person telling a lie – or even reacting to a question that would incriminate him or her – would exhibit changes in pulse rate, blood pressure, respiration, and the electrical conductivity of the skin. Its predecessor had been invented in 1895. Named the hydrosphygmograph, it measured variations in pulse rate and blood pressure of a subject under interrogation. The polygraph, introduced in 1921, made a continuous record of these and other changes.

At first, during the 1930s, the lie detector was believed to be infallible. However, in 1938, tests performed during the investigation of a kidnapping murder in Florida suggested that an innocent man was guilty and one who later confessed to the crime was innocent. J. Edgar Hoover, director of the FBI at that time, ordered his agents to "throw that box into Biscayne Bay." In 1964, he totally banned the technique.

Professor Udo Undeutsch of Cologne University explores the working of the polygraph, at the request of Germany's Federal Court of Justice, to determine whether its use could be permitted in criminal trials.

The interrogating detective seated himself in the dark behind a bright light blinding the eyes of the person being questioned. Other shadowy figures lurked in the background, coming and going at what the suspect recognized as critical points in the threatening interrogation. What had he just said? Had someone gone to check it out? His physical discomfort and mental unease frequently led to the required confession. This, however, is certainly the crudest of interrogation techniques – short of the

brutality of disorientation methods or torture – and its use has been discouraged.

Other methods continue to be used. In 1978, the FBI reinstated a small Polygraph Unit, and individual police authorities in the United States still make occasional use of the equipment. Investigators first attempt to calibrate an individual's response to undeniably factual questions and then, if possible, to deliberate lies. Following this, the individual is asked a series of innocuous questions interspersed with key questions directly related to the inquiry, and the reactions to the key questions are compared with those that represent deliberate lies.

Success with the polygraph depends on the individual's belief in its capabilities. When three American nuns and a lay worker were raped and murdered in El Salvador, the U.S. State Department demanded that Salvadorian National Guardsmen known to have been in the vicinity be tested. To demonstrate the powers of the equipment, an FBI agent told the men to write a number on a piece of paper and conceal it. When he successfully detected the numbers, four of the guardsmen were so impressed by his apparent powers that they immediately confessed and provided all the information that was required. Their fear of being found lying during subsequent

Pierce Brooks (left) and Robert Ressler, who cofounded the FBI's VICAP program in 1982. Brooks and Ressler – along with other FBI personnel – took part in hostage negotiation exercises.

Clifford Irving, the British man found guilty of forging the autobiography of millionaire Howard Hughes in 1971, with his wife. During inquiries into the source of the autobiography, Irving successfully passed a polygraph test.

interrogation proved a sufficient threat.

However, the shortcomings of the polygraph are well recognized, and U.S. authorities have published a list of types of individuals considered entirely unsuited to polygraph testing. These include psychopathic liars; emotionally or mentally disturbed persons; drunks; young children; and even those suffering from colds, asthma, or emphysema.

A striking example of the failure of the polygraph test is that of Clifford Irving, a man suspected – and later found guilty – of forging the autobiography of reclusive millionaire Howard Hughes in 1971. Irving was on his way to catch a flight from New York back to his home on Ibiza when he was asked to submit to a lie detector. He was emotionally exhausted and so preoccupied with worry that he might miss his flight that he passed the test. It was only after a check paid to "H. R. Hughes" was traced to the account of Irving's wife that the forgery was uncovered.

Recently, new instruments have been developed that can hopefully replace the polygraph. One, the Psychological Stress

Evaluator, can supposedly detect audible tremors in a suspect's voice – during interrogation, over a telephone, or in a tape recording – that will indicate whether he or she is lying. Such equipment could prove invaluable, not only during interrogation, but also in those cases where an UNSUB feels compelled to make telephone calls or send an anonymous tape recording to the police.

THE INTERROGATION OF DARRELL DEVIER

In December 1979, Mary Frances Stoner was reported missing in Rome, Georgia. A friendly, innocent 12-year-old, a drum-majorette in her school band, she had gotten off the bus at the driveway to her home. Her raped body was later found in a forest 10 miles away. Her head was covered with her coat; she had been struck several times with a large rock which lay, stained with blood, beside her. She had also been strangled manually from behind.

This was one of the earliest cases in which FBI profiler John Douglas was requested to give his advice, and it later became the basis of more sophisticated techniques. He rapidly assessed the situation and concluded that the perpetrator was a white male, in his mid to late 20s; he was probably a high school dropout, but with an above average IQ and a cocky personality. He had been discharged from

Police in Rome, Georgia, interrogated murder suspect Darrell Devier at night: FBI profiler John Douglas believed he would be vulnerable in this intimidating environment.

BODY LANGUAGE

The experienced interrogator knows that the human body can unconsciously reveal more than is being said. Recognizing this, the FBI has recently added instruction in body language to the training of agents. Robert Trott, head of the training program at Quantico, said: "Training includes deciphering all the clues you can get – not just what someone tells you in an interview, but all the signals they may give off."

Communication between two people confronting one another is based on more than words: the actual words used represent only 7 percent of the conversation. The volume, pitch, rhythm, and so on of the voice account for 38 percent. Facial expressions and body posture communicate 55 percent. And, in many cases, the body language visibly contradicts what is said.

For example, a person telling a lie may unconsciously raise a hand so that it begins to hide the mouth. One sitting with folded arms looking away from the speaker or with legs crossed at the ankles is shutting off the message that is being given. Posture can also provide indications of tension and anxiety, and the FBI believes that this can prove important in the questioning of suspected terrorists.

They quote a specific example. A man named Ziad Jarrah was stopped by a police officer in Maryland for speeding on September 9, 2001. He was given a ticket and a brief reprimand. Two days later the crumpled ticket was found in the glove compartment of a car parked at Newark airport soon after Jarrah and three others had hijacked the airliner that crashed in western Pennsylvania on the way to attack the White House.

While officials stress that the police officer had not failed in his duty, they believe that terrorists exhibit obvious signs of anxiety – such as the posture they adopt and the way in which they hold their hands – before an attack, and that law enforcement officers should be trained to detect them.

There are, however, drawbacks to the superficial aspects of body language analysis. Gestures in particular can have widely varying meanings in different cultures. In India, for example, shaking the head from side to side is not a sign of disagreement but of concentrated interest in the subject. Even looking at the other person with a steady gaze, which is commonly regarded as a sign of sincerity and truth, is far from a reliable indicator. Puerto Ricans consider it a sign of disrespect – and so a Puerto Rican who avoided looking his FBI interrogator in the eye would be considered shifty and untruthful, while actually being respectful.

military service on dishonorable or medical grounds, and had marriage problems, or was divorced. He was a blue collar worker, and he might have a previous criminal record for arson or rape.

Douglas reconstructed the crime situation in his imagination. The UNSUB almost certainly knew Mary Stoner by sight; he had coaxed her into his car, threatened her with a knife or gun, and driven off to an area where he knew he would not be disturbed. There was no leaf debris or dirt in her clothing, indicating that she had been made to undress before the rape, and then told to dress again. At this point, the UNSUB realized that the young girl could identify him; he had tried to strangle her to death, and finally killed her with the rock.

On the telephone to the Rome police, Douglas told them they had probably already interviewed the perpetrator. He would be cooperative and confident. The crime was likely not his first, but probably his first murder, and it had not been planned in advance. And his car, said Douglas, was black or blue; it would be some years old, because he could not afford a new one, but it would be well maintained.

The Rome police said that the profile exactly fit a man that they had questioned as a suspect. His name was Darrell Gene Devier, a 24-year-old white male who had been working as a tree-trimmer close to the Stoner home. A high school dropout, he nevertheless had an above-average IQ, rated at 100-110. Devier had been married

and divorced twice, and he was currently living with his first ex-wife. He had joined the Army after his first divorce, but was dishonorably discharged; and he was a suspect in the rape of another young girl, but had not been charged due to lack of evidence. His car was a well-maintained black Ford Pinto, three years old.

The police told Douglas that Devier was due for a polygraph test that day. Douglas advised against it, assuring them that the suspect would pass it and become more confident about any subsequent interrogation; however, the police went ahead, and then reported that the result was inconclusive.

"There's only one way to get to him," said Douglas. "Stage the interrogation at the police station at night." Devier would be more vulnerable to questioning in this kind of intimidating environment, and the local FBI agent should be present to add to the seriousness of the occasion. Stacks of file folders with Devier's name on them – even if they were filled with blank paper – should be on the desk. And – most importantly – the bloodstained rock should be on a low table, in such a position that Devier would have to turn his head to look at it. "If he is the killer, he will not be able to ignore it."

Douglas advised that the lighting be low, and that only one police officer and the FBI agent be in the room at any one time. What they had to imply was that they understood the stress that Devier was under. However repugnant they found the method, they should suggest that the

> **Most importantly, the bloodstained rock should be on a low table, in such a position that Devier would have to turn his head to look at it.**

victim had enticed him and then threatened to blackmail him. And they should tell him: "We know you got blood on you; on your hands, on your clothing. The question for us isn't 'Did you do it?' We know you did. The question is 'Why?' We think we know why, and we understand. All you have to do is tell us if we're right."

Douglas's advice was followed: Devier confessed to the murder, and to the other rape of which he was suspected. He was tried and convicted, and sentenced to the electric chair in Georgia. As this case demonstrates, profilers can successfully advise on cases where a polygraph test would be ineffective.

COLIN IRELAND: THE UNCOMMUNICATIVE SUSPECT

Interrogation can present problems when the suspect – either on the advice of his attorney or on his own initiative – declines to answer questions, with a brief "No comment," or even maintains a stolid silence. In his book *Picking up the Pieces* (2000), psychologist Paul Britton details the advice he gave to the British police questioning serial killer Colin Ireland.

On March 9, 1993, the newsroom of the British *Sun* tabloid newspaper received a telephone call. "I have murdered a man," a gruff London voice said, and gave an address in southwest London. "I'm calling you because I'm worried about his dogs, I want them to be let out. It would be cruel for them to be stuck there…. I tied him up and I killed him, and I cleaned up the flat afterwards. I did it. It was my New Year's resolution to kill a human being…. He was a homosexual and into kinky sex."

The police were informed and went to the given address. There, besides two dogs,

they found the naked body of 45-year-old theater director Peter Walker with his wrists and ankles knotted to each corner of his four-poster bed and a plastic bag tied tightly over his head. At first, detectives wondered whether Walker – a self-confessed homosexual – had died accidentally during a solitary sex practice, but it soon became evident that another person had been involved because he could not have tied himself to the bed and then secured the bag over his head. And the anonymous telephone call indicated that this was murder.

On May 28, 37-year-old librarian Christopher Dunn was discovered in almost identical circumstances in his home in northwest London – but this time local detectives decided that the death was accidental. It was not until after two more killings – of 35-year-old HIV-positive American executive Perry Bradley on June 4 and 33-year-old residential home supervisor Andrew Collier on June 7 – that police realized that the deaths were linked, when they received a series of telephone calls from the killer. He said that he had read the FBI's *Crime Classification Manual* (see Chapter 3), and had decided to become a serial killer. "I know what it takes to become one. You have to kill over four to qualify, don't you? I plan to kill five…."

Inquiries established that at least three of the victims were habitués of a well-known gay pub, The Coleherne, on Brompton Road in West London. Yet despite the gay community's fear of the man the press dubbed "The Fairy Liquidator," it was from The Coleherne that the fifth victim emerged on June 12, to encounter his killer at nearby Earls Court

Far left: Colin Ireland stalked and murdered five homosexual men in London in 1993. He told British police in a telephone call how he planned to become a serial killer: "I know what it takes to become one. You have to kill over four to qualify, don't you?" His fifth victim, seen here, was chef Emmanuel Spiteri.

subway station. He was 43-year-old chef Emmanuel Spiteri. The two men traveled to Charing Cross railroad station and on to Spiteri's apartment in southeast London. Spiteri's body was found there three days later, and a swift review of video surveillance films from Charing Cross revealed Spiteri accompanied by a tall, heavily built man. Police released his description and a computerized composite sketch, but after several weeks nobody had come forward to identify him.

Then, on July 21, a heavily built man named Colin Ireland walked into a lawyer's office in Southend-on-Sea, 30 miles from London. He said that he needed a lawyer; he admitted he had been with Spiteri on the night in question, but claimed that he had left when he found another man in the apartment. However, soon after he arrived with his attorney at a police station in London, his fingerprints were found to match one discovered in Andrew Collier's home. At this stage the police expected to obtain a ready confession with all the details they required for a successful conviction. But they did not realize the personality they were up against.

Ireland was born in 1954, and had a long career of petty crime behind him. He had served a prison sentence in 1981 for robbery and had spent some time in the army. He claimed to have been in the French Foreign Legion for a few months. He was reported to be proud of his reputation as a self-sufficient loner, having learned, he said, survival techniques in the army, and he would often spend nights by himself in the Essex marshes building bivouacs and snaring rabbits and birds for food.

Britton warned the police: "Every soldier knows to give nothing away under interrogation.... That's what he'll try to do." However: "He isn't truly a trained soldier. He doesn't have the personal resources to sustain himself indefinitely. It is extremely difficult to maintain a silence, and he doesn't yet know how difficult. He's going to learn."

Ireland, said Britton, believed that he was stronger, both physically and mentally, than his interrogators, and he regarded the coming interview as a contest of wills. It was essential that the police should never reveal frustration; they should be confident and relaxed and continue talking relentlessly. Even if it seemed to be a totally one-sided interview, Ireland would be listening – and listening for any sign of weakness or fatigue.

Ireland had been interviewed on tape two years previously, following a charge of actual bodily harm in a domestic dispute. Britton suggested that the police play this tape and point out that Ireland no longer had any need to remain silent because his voice could be identified as that of the anonymous caller who claimed to be the killer. "You want to provide him with every opportunity to give an account of himself, so there is no chance of a mistake being made, or of him being wrongly accused...."

If Ireland still refused to cooperate, said Britton, the officers should draw an

> Said Britton: "It is extremely difficult to maintain a silence, and he doesn't yet know how difficult. He's going to learn."

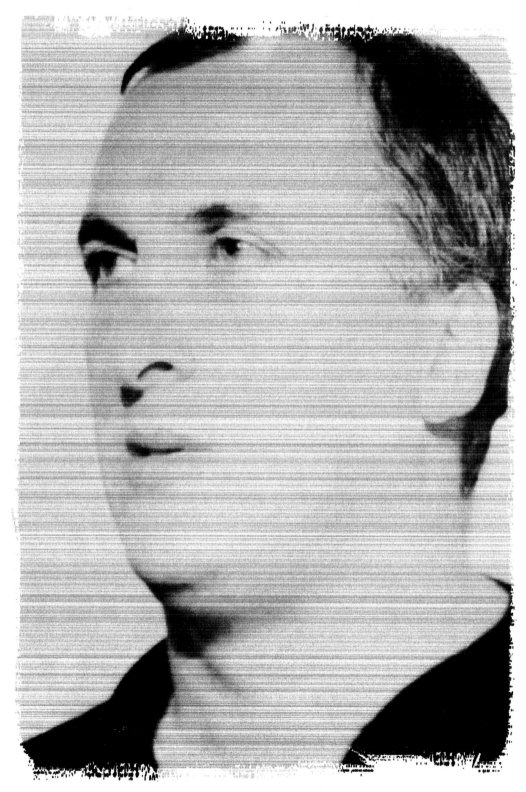

Proud of his reputation as a self-sufficient loner, Ireland claimed to be a trained soldier and believed that he would give nothing away under interrogation. Persistent questioning by the police, however, eventually weakened his resolve – although he did not finally confess until after three weeks in prison.

unflattering picture of him, questioning him on unsuccessful incidents in his past, such as his two failed marriages and other relationships. "He wants to be respected. If you paint him in a poor light, he will feel a powerful urge to correct you.... Don't challenge or confront him. That's what he expects. That's what he's steeling himself for."

But after two days, Ireland had still admitted nothing. He was remanded and moved to prison. Said Britton: "He will still be hearing the questions you ask him.... It won't be a question of using survivalist skills any more.... If he is to survive in any way he has to put this right. He has to set the record straight. That's when he'll confess." Three weeks later, Ireland asked to speak to the police – "but not those bastards who interviewed me. They really got under my skin." In a detailed statement that took two days to record, he confessed to all five murders, and provided a wealth of detail. At London's Old Bailey in December 1993, he was sentenced to five life sentences with a recommendation that he should never be released.

PSYCHOLOGICAL PRESSURE IN THE COURTROOM

When intensive interrogation has failed to produce a confession, it is still possible for psychological pressures to be applied, even during a trial. While Wayne Williams was being questioned in connection with the Atlanta child murders in 1981, he denied all responsibility and even agreed to take a "lie detector" test, which proved inconclusive. (Police and FBI agents who later searched his home found books that detailed ways to beat the test.) At his trial

he entered a plea of "not guilty," and was represented by a high-profile legal defense team. The problem that faced the prosecution was that Williams was well spoken and cooperative in his manner; he wore thick glasses and had delicate features. He lived at home with his parents, who were both retired schoolteachers, and he maintained that the accusations against him were purely racial in nature. The jury was very likely to be swayed by these factors, which did not fit in with the commonly held picture of a killer.

Because of his earlier advice on the case (see page 246), John Douglas was invited by the prosecution to attend the trial. For several days, the hearing did not go well for the prosecution. Despite their expectation that Williams would not testify, he announced that he would. His attorney kept him on the witness stand for more than a day, regularly making the same point to the jury: "Look at him! Does he look like a serial killer?.... Look how soft his hands are. Do you think he would have the strength to kill someone, to strangle someone with these hands?"

How was Williams to be cross-examined? Douglas advised assistant district attorney Jack Mallard: "You've got to keep him on the stand as long as you can – you've got to break him down. Because he's an over-controlled, rigid personality...and to get to that rigidity, you have to keep the pressure on him, sustain the tension by going through every aspect of his life...." Then, said Douglas, when Mallard had worn Williams down, he should physically touch him, moving in close, violating his space, and ask him in a low voice, "Did you panic, Wayne, when you killed those kids?"

WAYNE WILLIAMS

Williams protested his innocence as he left court after his arraignment, claiming that the accusation was racially motivated.

Police in Atlanta, Georgia, first knew that a killer of children was in the region when the bodies of two strangled boys, aged 13 and 14, were found in July 1979. In November, two more – one only nine years old – were murdered. By July 1980, the toll of dead and missing had reached 12. All the victims were black, and the local community began to fear that the killer was a member of a white racist group. The mayor of Atlanta appealed directly to the White House for the FBI to be brought into the case. Although the FBI does not usually intervene actively in murder cases, there were two plausible justifications. Since some of the children were still listed as "missing," they might plausibly have been kidnapped. And because the case could be racist, there was also the question of "violation of Federal civil rights."

By the time John Douglas and Roy Hazelwood arrived in Atlanta, the number of cases had risen to 16. Reviewing the case files, they both discounted the racist theory. The presence of a white murderer in the all-black neighborhood from which the children had been abducted would have been noticed immediately. Although the FBI agents suspected that two of the cases of young girls were not the work of the principal offender, they drew up a profile. They said that the murderer was black – in itself very unusual for a serial killer – single, aged 25–29. He would drive a police-type vehicle and own either a German shepherd or a Doberman. Without a girlfriend, he would be attracted to young boys, but there was no evidence that the boys had been sexually abused. "He would have some kind of ruse or con with these kids," Douglas wrote in *Mindhunter*. "I was betting on something to do with music or performing."

By March 1981, the death toll (including

children missing, presumed dead) stood at 26. The investigation was costing the city of Atlanta $250,000 a month, and President Reagan committed $1.5 million of Federal funds in assistance.

After two bodies were found floating in the Chattahoochee river during April, the police mounted surveillance on the principal bridge. During the evening of May 22, the crew of a patrol car heard a loud splash and saw a man driving off in a station wagon. When stopped by a second patrol car, the driver identified himself as Wayne Bertram Williams, age 23, a "music-biz talent scout."

At that time, Williams was only cautioned, as he said that he had been dumping trash in the river. But two days later, when the body of 27-year-old homosexual Nathaniel Cater was found in the water, Williams was arrested.

An eyewitness reported that Williams had been seen with Cater. Testimony was given later that Williams had made a number of homosexual advances to young men. At his trial in January 1982, he was charged with only the last two of his murders; telling forensic evidence – including hairs from his German shepherd found on the body of one of the victims – was produced by the FBI laboratory. He was found guilty and sentenced to life imprisonment. Although there were protests that he had been "railroaded" through unsupported circumstantial evidence, the Georgia Supreme Court upheld the verdict. And one thing is certain: there were no further serial murders of children and young men in Atlanta in the months following Williams's arrest.

Police assemble for the grim task of searching for clues where the body of 13-year-old Curtis Walker was found. Walker's was the 20th murder attributed to Williams.

And so, after several hours of persistent questioning, Mallard did just that. "No," replied Williams in a weak voice – and then he flew into a rage. He pointed at Douglas sitting behind the prosecution's table and shouted, "You're trying your best to make me fit that FBI profile, and I'm not going to help you do it!" He raved on, calling the FBI "goons" and the prosecution's team "fools." And that was when the jury realized for the first time that a violent side to Williams's nature had been kept hidden from them. It was the turning point of the trial.

This case, and many others, illustrates the importance of the psychological assessment techniques that have been developed over the past 25 years by the FBI and other law enforcement agencies throughout the world. Although considerable criticism has been leveled at what some consider an unjustified reliance upon the value of psychological profiling, this approach is being constantly developed and refined. The real-life Sherlock Holmes of the future will no doubt find it an increasingly vital tool in the investigation and solving of crime.

Wayne Williams had been questioned several times by Atlanta police before he was arrested on June 21, 1981.

Williams speaks at a news conference in jail, following his conviction as the Atlanta serial killer. He appealed, but the Georgia Supreme Court upheld the original verdict.

BIBLIOGRAPHY

Ainsworth, Peter B., *Offender Profiling and Crime Analysis*. Portland, Oregon: Willan Publishing, 2001.

Bartol, C., *Criminal Behavior: A Psychosocial Approach*. Toronto: Prentice-Hall, 1991.

Blackburn, Ronald, *The Psychology of Criminal Conduct*. New York: John Wiley & Sons, 1993.

Bluglass, R. & P. Bowden (eds), *Principles and Practice of Forensic Psychiatry*. London: Churchill Livingstone, 1990.

Britton, Paul, *The Jigsaw Man*. London: Transworld, 1997.

___. *Picking up the Pieces*. London: Bantam Press, 2000.

Brussel, James, *Casebook of a Crime Psychiatrist*. New York: Simon & Schuster, 1968.

Canter, David, *Criminal Shadows*. London: HarperCollins, 1994.

___. & Laurence Alison (eds), *Profiling in Policy and Practice*. Brookfield, Vermont: Ashgate, 1999.

Cook, Stephen, *The Real Cracker*. London: Fourth Estate, 2001.

Douglas, John, & Mark Olshaker, *Mindhunter*. New York: Scribner, 1995.

___. *Journey into Darkness*. London: Heinemann, 1997.

___. *Obsession*. New York: Scribner, 1998.

___. *The Anatomy of Motive*. New York: Scribner, 1999.

Foster, Don, *Author Unknown*. New York: Henry Holt & Company, 2000.

Green, E., *The Intent to Kill*. Baltimore: Clevedon, 1993.

Holmes, Ronald M., *Profiling Violent Crimes*. Newbury Park, California: Sage Publications, 1989.

Jackson, Janet L. & Debra A. Bekerian (eds), *Offender Profiling*. New York: John Wiley & Sons, 1997.

Jeffers, H. Paul, *Who Killed Precious?* Boston: Little, Brown, 1991.

Kind, Stuart, *The Scientific Investigation of Crime*. Harrogate, England: Forensic Science Services, 1987.

Langer, Walter, *The Mind of Adolf Hitler*. New York: New American Library, 1972.

Leyton, Elliott, *Compulsive Killers*. New York: New York University Press, 1986.

___. *Men of Blood*. London: Constable, 1995.

Lowe, Sheila, *Handwriting of the Famous and Infamous*. New York: Metro Books, 2001.

Lyman, M.D., *Criminal Investigations: The Art and the Science*. Englewood Cliffs, New Jersey: Prentice Hall, 1993.

Markman, Ronald, & Dominick Bosco, *Alone with the Devil*. Boston: Little, Brown, 1989.

Maren, Patricia: *The Criminal Hand.* London: Sphere, 1991.

Norris, Joel, *Serial Killers.* London: Arrow, 1988.

___. *Walking Time Bombs.* New York: Bantam Books, 1992.

Raskin, D. C. (ed), *Psychological Methods in Criminal Investigation and Evidence.* New York: Springer, 1989.

Ressler, Robert K. & Tom Schachtman, *Whoever Fights Monsters.* New York: Simon & Schuster, 1992.

___. John E. Douglas, Ann W. Burgess & Allen G. Burgess, *Crime Classification Manual.* New York: Lexington Books, 1992.

___. *Sexual Homicide.* New York: Lexington Books, 1988.

Rhodes, Richard, *Why They Kill.* New York: Knopf, 1999.

Rossmo, D. Kim, *Geographic Profiling.* Boca Raton, Florida: CRC Press, 2000.

Star, J. & J. Estes, *Geographic Information Systems.* Toronto: Prentice-Hall, 1990.

Turvey, Brent, *Criminal Profiling.* San Diego: Academic Press, 1999.

Williams, Katherine S., *Textbook of Criminology.* New York: Oxford University Press, 2001.

Wilson, Colin, *Written in Blood.* London: Grafton Books, 1990.

___. & Donald Seaman, *The Serial Killers.* London: Virgin Books, 1996.

JOURNALS

American Journal of Psychiatry
American Sociological Review
Behavioral Sciences and the Law
British Journal of Criminology
Bulletin of the American Academy of Criminology and Law
Criminal Justice and Behavior
FBI Law Enforcement Bulletin
Journal of Abnormal Psychology
Journal of Contemporary Criminal Justice
Journal of Criminal Law and Criminology
Journal of Forensic Psychology
Journal of the Forensic Science Society
Journal of Interpersonal Violence
Journal of Police Science and Administration
Police Review
Police Studies
Psychology, Crime and Law
RCMP Gazette
Science & Justice
Science & the Law
Studies on Crime and Crime Prevention

WEBSITES

The following websites provide useful information on criminal profiling:

www.corpus-delicti.com/profile.html
www.crimelibrary.com
www.criminalprofiler.com
www.criminalprofiling.com
www.fbi.gov/publications/leb
www.forensic-crim.com
www.ramas.co.uk/offender
www.wm3.org/html/profile.html

INDEX

Page numbers in *italics* refer to illustrations

PICTURE CREDITS